Fishery Technology

Development Lessons for the Third World

Published by

S.B. Nangia
for **APH PUBLISHING CORPORATION**
4435-36/7, Ansari Road,
Darya Ganj,
New Delhi-110 002
Tel. 23274050

Fax 011-23274050
Email: aphbooks@gmail.com

2025

Printed in India at
Balaji Offset
New Delhi-110 035

Fishery Technology

Development Lessons for the Third World

P. Suresh Kumar

APH PUBLISHING CORPORATION
4435-36/7, ANSARI ROAD, DARYA GANJ,
NEW DELHI-110 002

Preface

The present work 'Fishery Technology: Development Lessons for the Third World' is an outcome of a dissertation submitted by the author as a part of Ph.D Programme in the Cochin University of Science and Technology (CUSAT) in the year 1999. Though about a decade has passed since its submission, the justification for its present publication primarily stems from the profound relevance of the analytical contents of the study in the context of post – globalization scenario in India, and the Third World at large.

The Third World economies, by and large, were concerned with growth and development at much a formal level since the Second World War. They were thrown into an urgency of development in the context of their newly won political freedom. Apart from that, the level of material progress achieved by the colonial powers enticed the poverty stricken areas in the world to aspire for economic development of a tall order. Their newly found enthusiasm of development and progress had impelled them to emulate and replicate various models, programmes and policies as modernalization packages, exclusively built for them by the intelligentia – academia of the leading developed nations. However, sooner or later, such attempts ended up in failure. The experiences proved the invalidity and inapplicability of prescribing homogenous and universal laws for growth and development.

The global division of the world economy into a developed one and a less developed ome had emerged during the colonial phase of global economic transformation. This 'development gap' could not be bridged by the facilitation of the imported variety of modernization attempts. More over, the implementation of such 'development paradigms' had recreated and intensified the economic cleavage in the global arena.

The colonial savageries that have scarred the psyche of these economies made them myopic in their development vision. The perceptions on development germinated from such a milieu have deprived them from realizing the strength of their historically evolved value systems and organizational or institutional powers and its credibility. The organic and nascent frames never became conjunct in their re-construction efforts. Rather, all such finer dimensions were replaced with seemingly more objective variables and parameters that had enormously strengthened the colonial powers during their colonial reign. The faltering of development efforts of the developing world occurred mainly from discarding the inherent and internal strength and its local wisdom.

A relatively independent phase in the global economy had emerged when the forces of production such as capital, technology and a lot of institutional mechanisms were freed from the clutches of the capitalist domination that had entrenched the less developed economies since the time of colonization. The fluidity and flexibility of the forces of production under the new epoch of globailisation has produced potential opportunities to the Third World. Some of them have succeeded to appropriate such positive realms by harmonizing the new situations with their internal dynamics securing improved growth rates and development indices.

However, the capitalists belong to such a clan that their intrigues gave them a phoenix power to regain their domination over the global economic environment. Their conscious efforts in turning the new situation to their own advantage succeeded in numerous ways. This included the manipulation of the supposed democratic global institutions such as IMF, World Bank and WTO. The global rules of economic transactions ordered in the global arena shortly deprived the Less Developed Economies of their opportunities of self – determined and self initiated growth process. Thus the so-called neo-liberal policies that had emerged in the context of globalization proved itself to be baneful to the developing world. Such policies neither were people centered nor took care of the livelihood issues of common people. The orchestrated policies and programmes only restored and strengthened the historical economic subjugation of the less developed economies to the mighty capitalist powers.

The present phase of globalization, however, has brought out a qualitative shift in the perceptions of development among the global community. The incressing marginalization of large sections of people and the loss of livelihood of the general mass as a continuum of capitalist development process are being increasingly realized by the global civil society. Members of the civil society such as non-governmental organizations (NGOs), philanthropic institutions, voluntary groups, U N agencies, elite academia, human rights groups and a host of similar organizations assert the rights of men to have a decent existence. These agencies reckon the short comings and weaknesses of the conventional development strategies as thwarting the basic human survival concerns.

The present work narrates the story of the marginalization process meted out to an original community of fishermen in Kerala in the guise of modernization based on capitalist paradigms of development. Their resistance and attempts of retrieving their work and labour process are lessons worth considering in the context of extensive marginalization meted out to similar communities around the world due to the onslaught of globalization. In the context of the widening base of democracy, the universal acceptance of the principle of basic empowerment of the under privileged and the emergence of a highly sensitized civil society, it is hoped that this case study, would contribute immensely to the onward movement of the human society with more dignity and to greater material progress.

The study as a Ph.D programme was supervised by the eminent international scholar and the former World Bank consultant Prof. (Dr). D.Rajasenan, currently working as Director, International Centre for Economic Studies and Policy Analysis, CUSAT. The services rendered by this peer teacher is duly acknowledged with high reverence. An equally important name that I should revere in this regard is that of Prof.(Dr)KC. Sankaranarayanan , Former Head of the Department of Applied Economics, CUSAT.

This work has come to its present form due to the constant inspiration of my friend Dr.R.K.Suresh Kumar, Head of the Department of Political Science, University College, Thiruvananthapuram. I am also thankful to Dr.P. Sukumaran Nair, Head of the Department of Political Science, N.S.S.College, Pandalam for his valuable advices in connection with the publication of this work.

P. Suresh Kumar

Contents

List of Tables

List of Figures

1
Introduction

One major concern of Economics is to secure economic prosperity. To Adam Smith, economic prosperity depends on capital accumulation and productivity (Dasgupta, 1985). Capital-rich economies have made substantial improvements in labour productivity and capital has instrumental in transforming them into developed economies. This elegant transformation process was projected as model for the development of new aspirants: the under-developed (Third) world. Such development ideas and strategies for Less Developed Countries (LDCs) came to be called development economics or modernisation theory and the economists of the persuasion as 'modernisation theorists'. They believed that replication of the same set of conditions of the developed world would usher in development in LDCs as well. Rostow's 'stages of growth' theory required all nations to traverse the same path[1]. The modernisation theory contended that there were certain 'missing components' in LDCs such as capital, foreign exchange and skills of management and that the presence of these would generate development in the Third world (Rostow, 1960). However, in due course of time it became clear that the development problems of LDCs were too complex to be solved through a simple linear development approach[2].

Marxian theory was also concerned about such issues and viewed development as problems of social relations. It held that the development of capitalism was auto-centric and transitory. Marxian theory held that a native land in touch with a capitalist mode would also be transformed so. Marx did not deal with the relationships between capitalist economy and a native LDC. It was later-Marxian thinkers who were engaged in with some reflections on the subject. Imperialistic stages of development as conceived by Lenin provide the general contours of the relation between capitalist countries and the less developed ones (Lenin, 1964)[3]. These stages of capitalism introduce contradictions into LDCs through imperialistic measures. This piece of analysis has not qualitatively changed the Marxian approach towards development issues of LDCs.

A flaw in the above strands of thought is the premiss that the internal structure and forces that cause changes in the LDCs are universal and homogeneous. Such theories failed to recognise that the developed countries and LDCs stand apart requiring different strategies and approaches to solve their respective issues. The concept of universal and homogeneous structure in the LDCs implies a static and ahistorical relationship between capitalist economies and LDCs.

Marxist structuralists, popularly known as Dependency School[4] upheld the view that the LDCs are often deprived of an independent growth path by a coalescence of domestic and foreign capital. Furtado (Furtado, 1963) cites the example of Brazil where foreign and domestic capital colluded to preserve the basic structure of the economy unchanged for the purpose of exploiting the reserve army of unemployed. To him such basic structural contradictions thwart developmental efforts.

The causes of such structural imbalances were well articulated by Frank (Frank, 1975). While analysing the capitalist underdevelopment, he argued that development and under development were two sides of the same historical process of world capitalist development. He argues that "under development as we know it to-day, and economic development as well, are the simultaneous and related products of the development on a world wide scale and over a history of more than four centuries at least of a single integrated system capitalism" (Frank, 1975). Frank rightly contends that capitalism is an alien system in LDCs. In LDCs it was not a domestic development as it was in Europe but the result of external imposition and control. The compensatory development of viable institutions and forms did not take place in the colonial periphery. Capital was extracted particularly from extractive industries and siphoned off to the development of metropole undermining growth prospects of domestic commerce and industry. The institutions and attitudes that were created to replace the pre-existing feudal ones were not those which stimulated or permitted autonomous growth. It had resulted only in perpetuating and deepening dependent relationship. The formation of underdevelopment relationship is not specific to capitalist epoch only. The breeds of under development sprouted even from an earlier epoch of mercantilism. The development of mercantilism and of capitalism must be seen as a single continuous process culminating in under-development. For LDCs, capitalism to-day is very much the same as it was mercantilism centuries ago. Frank tells, "For the under developed part of the capitalist system, relatively little of importance has changed since mercantilist times" (Frank, 1975). In the metropole the transformation of the mercantilist system to its capitalist phase was accompanied by industrialisation. But it was not so in the underdeveloped periphery. The metropole, during the capitalist phase employed the mercantilist relations to strengthen industrialisation at home. Further, it employed the same relations to thwart industrialisation in the underdeveloped world. In short, metropolitan industrialisation only strengthened mercantile determination of underdevelopment.

Thus we may summarise Frank's position as that the development and under development are of dialectical unity as development of metropolis causes the under development of the other. The relative position of the developed and the underdeveloped ones at present are the outcome of this historical process. In fact, the LDCs are entrenched in the world capitalist system in such a way that it creates and sustains a whole "structure of under development" deepening, widening and reinforcing that structure through the manoeuvring capitalist measures.

Different writers who focused on development issues of LDCs from this perspective have pointed out certain explicit underdevelopment structures (Boeke, 1953, Stewart, 1977). According to Stewart, "The dependent relationship is exhibited in cultural as well as economic features of the third world countries" (Stewart, 1977). She puts that the most critical aspect of the whole set of dependent relationships is technological dependence[5]. She argues,

"It is linked as a cause, a symptom and a consequence of general relationship. Indeed, it is possible to argue that technological dependence is the most critical aspect of the whole relationship - so long as it continues it is impossible to break out of the general relationship and if it could be avoided then genuine general independence would be possible" (Stewart, 1977).

Technological dependence arises from the imbalances in technological capacity, i.e. the capacity to produce technology [6]. History posited the developed countries at a vantage point. Being pioneers in industrial development they could exploit the LDCs of their resources not only through unequal trade but also by exporting embodied technology when development came to be identified as industrialisation of the Western type. In fact, commercialisation of technology modulated by the export of machinery to LDCs became a general feature.

Stewart has argued that it is impossible for the LDCs to come out of this shackle. Further, the psychological and economic pressures of the dependent relationship have conditioned decision makers in the Third world countries in such a way that they do not wish to follow an alternative strategy. Hence the development premises of LDCs are normally fashioned after the development paradigms of the developed capitalist economies. In this milieu the technological dependence perpetuates itself as a vicious circle and permeates and gets transform into a general dependence.

As a whole, the above analyses show that the LDCs are cornered in a peculiar adverse situation, where:

(a) LDCs are deprived of an independent growth path

(b) Theories under different strands of thought offer only a unilinear approach of development for LDCs, incapable of giving real solutions to their development issues.

(c) A set of structural imbalances exist in LDCs due to their current inevitable relations with the capitalist world.

Given these situations of LDCs, naturally one may ask the question whether there is any way out from this dead lock ? Luckily we are not ensnared altogether. Discrete attempts are found in different sectors among a particular class of people, the marginalised and peripheralised group, who bear the major brunt of capitalist development. Even though such attempts are not on a large scale, to be very conspicuous, it occur in certain sectors where such attempts are spreading. One may also note that such attempts by people are not strictly confining to economic issues alone but entail at political and social level (Satyamoorthy, 1998).

This book is about such an independent attempt made by the coastal marine fishermen of the Kerala economy against the capitalistic entrenchment of the fishing sector. Fishing is an important economic activity in Kerala since time immemorial. It was a traditional occupation and under the impulse of modernisation, fishery development schemes and policies were framed in Kerala both by central and state governments. The most important attempt for fisheries expansion was undertaken with foreign participation in 1953 under the Indo - Norwegian Project (INP). A decade of planned policies came to a close by 1963. Since then the fishery sector has been dominated by capitalist expansion and as a result the whole marine fishing scenario has remained turbulent. The original target group, the real fishermen were thrown out of scene, marginalised and peripheralised depriving them even of a precarious existence. From this unexpected and undesired scenario of the Kerala fishery we are bound to ask some questions. First, how an expansion of fishery which was originally planned to improve the economic condition of the real fishermen resulted in depriving their economic existence?

Second, how a sector which was dominated by traditional fishermen suddenly got replaced by a modern capitalist group who were totally unrelated to the fishery sector earlier?

Finally, what would be the nature of the new socio-economic scenario developed in this sector since this epoch? From 1980s onwards, the traditional fishermen as a whole began to make a recovery through some improvements in their artifacts – which was transformed as a dominant epoch known as motorisation in Kerala fishery. The major issues related with this particular epoch are the kind of forces which caused this motorization, its impact and how this epoch was dialectically linked with the earlier capitalist epoch.

The general tone of all these issues are related with technology and dependency and the consequent structural imbalances. The intention here is to unfold all these issues at a broader developmental perspective in the context of fishery sector of Kerala.

Survey of Literature

Large number of studies have emerged in this area dealing with different phases of fishery development in the state. On the eve of Indo-Norwegian Project (INP), many studies were conducted particularly on the impact of INP on fishermen (Bog, 1959, Sandven, 1959, Achari and Menon, 1963, Klausen, 1968, Achari, 1969). All these studies have pointed out that generally INP provided an improvement in the economic well being of the fishermen. But a major limitation of these studies was that they reflected the economic condition of the INP area only.

In the 1960s and 1970s fisheries sector was proliferated with works on the experiences of mechanisation. Most of the studies were primarily concerned with technological issues such as evaluating comparative efficiency of different mechanised crafts (Iyer et. al., 1968: State Planning Board, 1969) and the externalities of such technology (Gaining, 1974).

A major change that occurred in the fishing sector of the state during the mechanisation period was the primacy accorded to shrimp exports. An economic appraisal of the prawn fishery of the Kerala coast was attempted by Saxena (Saxena, 1970). Valsala made a study on the structure of the marine product export industry (Valsala, 1976). Prakasan dealt with the impact of mechanisation on the fishermen of Vypeen Island (Prakasan, 1974). Vattamattom tried to identify factors that determine the earnings of the fishermen of the Poonthura Village in Trivandrum district (Vattamattom, 1978). All these studies aimed at analysing certain particular aspect of the mechanisation period.

The studies focusing on the general picture of the evolving pattern of the Kerala fishery were limited. However, attempts were made in this line mainly by writers like Kurien, (1978); Mathur, (1978); and Bhusan, (1979). Kurien (1978), focused on fisheries in terms of production, distribution, technology and organisation. A similar study of the Mappila Fisher Folk Mathur, (1978) relating to organisation, technology, trade and other related aspects in hindsight gave an insight to the whole fishery economy of the Kerala State. A detailed description of the technical changes in Kerala fishery was made by Bhushan (1979). All these studies enriched information relating to fishery but only at a descriptive level.

Explanatory attempts relating to the changes in Kerala fishery were made by Ibrahim, (1986); Ramakrishnan, (1994); and Kurien, 1996); Ibrahim, (1986) focused on the capitalist intrusion in the primary fishing activity and analysed its implications on employment and income. He has shown that the mechanisation process has adversely affected the income, and employment prospects of traditional fishermen. However, his study has not focused on how the capitalists succeeded in intruding into the primary sector and dominating it. Though he had used Marxian tools to delineate the capitalist changes, he left unfinished it in analyzing the changes in the labour process in the fishery sector.

Ramakrishnan (1994), highlighted the need for an appropriate theoretical framework to address and comprehend the process of fisheries development. He suggested an alternative approach to study technological change in the primary marine fishing industry of Kerala by referring to its process, indicators and characteristics. His failure to recognise the capitalist process of growth under the guise of modernisation of the sector resulted in the study ending up at a descriptive level. His basic premise, perhaps, not helped him to comprehend the motorisation phase of fishery development.

Kurien (1996), attempted to explore the diffusion of technology among the fishermen. Even though he had analysed the process of technology diffusion in terms of theoretical assertions, he could not bring the dialectical relation of the diffusion phase with the prior changes in fishery.

Fisheries literature in the 1980s mainly focused on the uncertainities and anxieties of different sections involved. The traditional fishermen were marginalised with the onslaught of mechanisation and they were almost deprived of their source of living. Such a situation created a conflicting relation between traditional fisher folk and the capitalists. Issues of over-exploitation and resource conservation were brought to the forefront (Kurien, 1987). The conflicting nature of certain technologies like purse-seining were discussed by Achari (1986). The impact of motorisation in the fisheries on traditional fishermen secured the attention of some experts, however, marginally (Kurien and Jayakumar, 1980, Achari, 1986). Costs and benefits of motorisation from the angle of techno-economic issues were studied by certain NGOs (PCO, 1989).

Even though all these studies touched upon the live issues of Kerala fishery a major shortcoming was that they could not visualize that such issues were organically brewed and emerged from the development paradigms adopted in the fishery sector. The problems highlighted in the fisheries are not isolated issues in fisheries alone. They are on the other hand, issues emerging in any traditional sector/economy opened for modernisation on western lines. Therefore, to analyse the evolutionary process of fishery and to comprehend the issues as the inseparable and inevitable outcome of capitalist development paradigms, we have to look at these issues from a critical development perspective. The survey of fishery studies indicates that such a gap exists in the whole literature (recall Ramakrishnan's attempt to formulate such a theoretical net work in terms of process, indicators and characteristics). The present study is an attempt to undertake this task, however marginally.

Given that, the crisis in Kerala fishery is the inevitable outcome of capitalist development paradigms, we would like to specifically examine the following issues.

Objectives of the Study

The objectives of the study are

1. To unfold the capitalist development process in the fishery sector.

2. To explain the marginalisation of the traditional fishermen in the development process of the fishery sector.
3. To explain and assess the responses of the fishermen community against their marginalisation.
4. To highlight the implications of such responses in the fishery sector of Kerala.

Hypotheses

The study makes following the hypotheses.

1. Export orientation of the fishery initiated capitalist development.
2. Motorisation seems to be an appropriate technology for traditional sector and its development.
3. Motorisation itself produces newer trends of competition in traditional fishery and even in the mechanised sector.

Methodology

Since the issues raised in this study are multidimensional, Marxian dialectical method forms the basis of analysis. The concept of labour process changes constitutes the central plank of the analysis[7]. Since the study is tailored as a case study, both primary and secondary data are used.

To collect the primary data, the sample size is determined on the basis of a census of artisanal craft made by South Indian Federation of Fishermen Society (SIFFS) in 1991. The samples are drawn from six coastal districts, three from the Southern region and the other three from the Northern region. On the basis of the types of crafts, a five-fold classification of crafts are made in both regions. From each region, five per cent of the population is drawn as sample from the important fishing villages. A questionnaire is used to elicit information from the fishermen. Basic statistical tools are used to analyse and explain data.

Scheme of Study

This study is divided into nine chapters. Chapter one, which has been set out as introduction, indulges in a brief discussion of some major alternative development theories and brings out their limitations in the socio-economic milieu of LDCs. The irrelevance and inappropriateness of such models which had culminated in serious deprivation of livelihood, particularly of primary producers have caused newer development visions and actions undertaken by people themselves. Given this context, the fishery sector in Kerala is taken as a case study. A brief survey of literature followed by the objectives of the study and methodology are discussed in this chapter.

The second attempts to make a theoretical explanation of the newer development trends which are emanating from the primary producers. The study adopts Marxian dialectics as its methodology particularly the labour process analysis. Having defined what is labour process, the study proceeds to explain how it looks into the pre-capitalist and capitalist modes of production. Then we describe the changes in labour process under different epochs of capitalist development. We contend later that changes in labour process takes unusual form in LDCs because of its peculiar of socio-economic environment. We also attempt to provide the reasons for newer forms of development activities by the marginalised community in this chapter.

While chapter three gives a simple narration of the evolution of technology in fishery, the fourth and fifth chapters show how in the fishery such technical changes culminated in class polarisation. It further discusses how the fishermen community reacted against such polarisation.

The epoch of motorisation which burgeoned as a reactionary measure turned up as an alternative development attempt. This story is described in chapters six and seven. The socio-economic viability of this epoch and its characteristics are tested with primary data in these chapters.

The role played by the government in the evolution of Kerala fishery is discussed in chapter eight. Policy directions to be followed in its future development are also given on the basis of the present analysis.

The final chapter (chapter nine) makes a summary of the study and its major conclusions.

NOTES

1. Rostow argued that an economy in its development process must pass through five stages. From a traditional stage in the begining to pre conditions as the second, which lay the ground work for a take off in the third phase. It progresses then into a drive to maturity and finally the fifth stage of high mass consumption (Rostow, 1960).
2. Referring to such modernisation approaches, Louis Leferber argues "that it is not applicable because it does not relate to those structural conditions which are present in today's

underdeveloped nations, and that it does not leave room for the attainment of social justice without which growth cannot be turned into development" (Leferber, 1974).

3. Lenin has described five essential features of imperialism.
   (i) The concentration of production and capital developed in to such a high stage that it created monopolies which play a decisive role in economic life.
   (ii) The merging of bank capital with industrial capital and the creation, on the basis of this, finance capital or a financial oligarchy.
   (iii) The export of capital, as distinguished from the export of commodities, becomes of particularly great importance.
   (iv) International monopoly combines of capitalists are formed which divide up the world.
   (v) The territorial division of the world by the greatest capitalist powers.
4. Celso Furtado, A.G.Frank, Sunkel, Dos Santos Szentes, Samir Amin, Griffin were the major exponents of this school of thought.
5. The dependent relationship means that events in the third world countries are determined by what happens elsewhere, notably at the capitalist centre.
6. Technological dependence, its characteristics and indicators and consequences are discussed by Frances Stewart (Stewart, 1977).
7. Chapter 2 describes the concept of labour process, labour process changes and specific issues related with labour process changes in LDCs.

REFERENCES

1. Achari, T.R.T, (1969): *The Impact of Indo - Norwegian Project on the Growth and Development of Indian Fisheries*, State Planning Board, Trivandrum.
2. Achari, T.R.T., and Menon, Devidas.M.,(1963): *A Report on the Assessment of the Indo-Norwegian Project on the Socio- Economic Conditions of the Fishermen of the Indo-Norwegian Project Area*, NORAD, Oslo.
3. Bhushan, Bharat; (1979): *Technological Change in Fishing in Kerala* 1953-77, (Unpublished M.Phil Dissertation),Centre for Development Studies,Trivandrum.

4. Boeke, J.H.,(1953): *Economics and Economic Policy of Dual Societies*, New York.

5. Bog, P., (1959): *A Statistical Survey of Economic Conditions in Project Area - Indo Norwegian Project Report No 2*. The Norwegian. Foundation for Assistance to Under Developed Countries, Oslo, 1959.

6. Dasguptha, A.K.,(1985): *Epochs of Economic Theory*. Oxford Press, Delhi.

7. Frank, Andre, Gunder., (1975) : *On Capitalist Underdevelopment*. Oxford University Press, Bombay.

8. Furtado, Celso., (1963): *The Economic Growth of Brazil: A Survey from Colonial to Modern Times*. Berkeley.

9. Gaining, John., (1974): "Technology and Dependence - The Internal Logic of Excessive Modernisation in a Fisheries Project in Kerala", CERES, September, October.

10. Ibrahim, P., (1986): *Fishing Industry in Kerala, A Study of the Impact of Technological Change* (Ph. D.Dissertation), University of Kerala, Trivandrum.

11. Iyer, H. Krishna, *et al*., (1968) : "Comparative Fishing Ability and Economic Efficiency of Mechanised Trawlers Operating Along the Kerala Coast", *Fishery Technology*. Vol. V, No.2.

12. John,Valsala.,(1976): *The Marine Products Export Industry of Kerala - Some Aspects of its Structure and Backward Linkages*. (Unpublished M. Phil Dissertation).Centre for Development Studies, Trivandrum.

13. Klausen, A.M.,(1968): *Kerala Fishermen and the Indo-Norwegian Project*. George Alien and Unwin, London.

14. Korakandy, Ramakrishnan.,(1994): *Technological Change and the Development of Marine Fishing Industry in India*. Daya Publishing House, New Delhi.

15. Kurien, John., (1978): *Towards an Understanding of the Fish Economy of Kerala State*, (Working Paper No.68), Centre for Development Studies, Trivandrum.

16. Kurien, John., (1987): "The Impact of New Technology Introduction into Fishing in Kerala", *Kerala Seminar on Fisheries Crisis and Policy Approach*, Fisheries Research Cell, Programme for Community Organisation, Trivandrum.

17. Kurien, John., (1991): *Ruining the Commons and Responses of the Commoners. Coastal Overfishing and Fishermen's Actions in Kerala State, India*. Discussion Paper 23, UN Research Institute for Social Development.

18. Kurien John., (1994): *Technology Diffusion in Marine Fisheries, The Concrete Socio-Economic and Ecological Interaction: A Study of the Diffusion of Motorised Plywood Boats Along the Lower South - West Coast of India (Ph.D Thesis), Tata Institute of Social Sciences, Bombay.*

19. Kurienjohn, and Jayakumar. S.R.J., (1980): Motorisation of Traditional Canoes - *The Purakkad Experiment.* Fisheries Research Cell, PCO, Trivandrum, 1980.

20. Leferber, L., (1974): "On the Paradigm for Economic Develop ment", *World Development,* Vol.2. No. I.

21. Lenin, V.I., (1964): "Imperialism : The Highest Stage of Capitalism", *Collected Works.* Vol. 22., Progress Publishers, Moscow.

22. Mathur, P.R.G., (1978): *The Mappila Fisherfolk of Kerala,* Kerala historical Society, Trivandrum.

23. Paul, A. Baran., (1992): "On the Political Economy of Backward-ness", *in Political Economy of Development and Under development,* Wilber, K.C., and Jameson, P.K., Me Graw-Hill.

24. Prakasan, M.S., (1974): "Impact of Mechanisation on Fishermen", *Voluntary Action*, Vol. XVI, Nos. 2 and 3.

25. Programme for Community Organisation. (1989) : *Techno Economic Analysis of Motorisation of. Fishing Units Along the Lower South-West Coast of India,* Fisheries Research Cell. P.C.O.Trivandrum.

26. Rostow, W.W., (1960): *The Stages of Economic Growth,* Cambridge.

27. Sandven, P., (1959): *The Indo-Norwegian Project in Kerala,* Norwegian Foundation for Assistance to under Developed Countries, Oslo. 167.

28. Satyamoorthi., (1998): "Labour of Sisyphus, Feast of Barmecide, The Sentence and the Premise in Development Studies", *Economic and Political* Weekly, Vol. XXXIII.

29. Saxena, B.S., (1970): "Use of Economic Perimeters in the Rationale Exploitation of Deep-Sea Fisheries with Particular Reference to Shrimp Fishery of Kerala Coast".

30. State Planning Board, (1962): *Indian Sea foods,* Vol. III, No.I. *Comparative Efficiency of Fishing Crafts in Kerala.* Govt. of Kerala, Trivandrum.

31. Stewart, Frances., (1977): *Technology and Underdevelopment,* The Macmillan Press Ltd., London.

32. Vattamattom, Joseph., (1978): *Factors that Determine the Income of Fishermen - A Case Study of Poonthura Village in Trivandrum District,* (Unpublished M.Phil Dissertation), Centre for Development Studies, Trivandrum.

2

A Perspective of Technology and Labour Process in LDCs

In this chapter, we delineate the concept of labour process and its changes. We would show, there upon, the necessity of distinguishing, analytically the labour process changes in LDCs as distinct from capitalist economies. Further, our endeavor is to contend that the distinct labour process set up in LDCs provide scope and possibilities for alternative development formations in LDCs focusing on fishing industry in Kerala as a case study.

The foundation of labour process analysis was laid by Marx (Marx, 1978). The organic nature of production is that it involves a labour process. At an abstract level, labour process may be defined as an interaction between human beings and nature on the realm of production. Labour is a process in which both man and nature participate and man on his accord, starts, regulates, and controls the material reactions between himself and nature. He opposes himself to nature as one of her own forces, setting in motion arms and legs, head and hands, the natural forces of his body, in order to appropriate natures productions in a form to his own wants (Marx, 1978).

At the production level, there are three elements in the labour process. They are:

1. The personal or purposeful activity of man, i.e labour or work itself.
2. The subject of that work, and
3. The instruments of labour.

Collectively these basic elements are called as the means of production and the process which results in the production of some use-value is defined as the labour process (Miles, 1987).

In the labour process, the labour and the consequent production of use value has a subjective link. This emerges, when the labour transforms the object of labour in a pre-determined way, using the instruments of labour. The end product resulting from the labour process thus possesses attributes which had in the imagination of the labourer. In short, the end product manifests the involvement of the entire spirits and aspirations of the labourer.

"At the end of every labour - process, we get a result that already existed in the imagination of the labourer at its commencement. He not only effects a change of form in the material on which he works, but he also realises a purpose of his own that gives the law to his modus operandi, and to which he must subordinate his will. And this subordination is no mere momentary act" (Marx, 1978).

Labour process, considered at the level of man - nature interaction is an independent and organic act. Also, it is a universal phenomenon. It is the everlasting nature - imposed condition of human existence or rather is common to every such phase (Marx, 1978). In other words, labour process involves a universal material realm. We cannot strip off the material content of the labour process. The material content of the production process will have a social form, for the content cannot exist without its form. This form bears the imprints of the mode of production within which the given labour process has evolved. In short, the labour process in existence at a particular stage of history bears the imprints of that historical stage and its mode of production. Thus we see an organic link between the labour process and the mode of production in the form of mutual interaction between them. To give a proper focus to the analysis of economic issues, a good methodology would be to give consideration to the dialectical relationship between labour process and mode of production[1].

Labour Process under Pre-Capitalist Mode of Production

Mode of production reflects the outcome of interaction of productive forces and relations of production. The former represents the material content and the latter the social form of development of the productive forces which is inseparably bound to the content. The nature and the level of development of productive forces determine economic relations among people, particularly the type of ownership of the means of production. The relations of production can promote or thwart rapid development and cause in certain situation, partial deterioration of productive forces. It is the dialectical relations between forces of production and relations of production that cause class cleavages and subsequent transformation into new progressive modes of production (Volkov, 1985).

The pre-capitalist mode of production was generally characterised by :

1 Use of simple tools and techniques,
2 Use of co-operative labour,
3 Subsistence rather than surplus making motive of production,
4 Lower productivity and
5 More or less collective ownership and sharing.

Under such material and social relations the labour process was such that the work remained under the control of real producers. Traditional knowledge has been embodied as skill among workers in their respective crafts. The work organisations under pre-capitalist mode involved the distribution of tasks, crafts or specialities of production throughout the society. It was not a systematic subdivision of the work into limited operations to call it as division of labour (Braverman, 1979). It is, to use Marx's terminology, the social division of labour. In this labour process men or women may habitually be connected with the making of certain products, but they do not as a rule divide up the separate operations involved in the making of each product. In other words, the labour process was such that no labourer remained separated from the work involved in the production and all had accessibility to any work activity since all were familiar with the activities involved in the work even though some of them did an activity regularly. More precisely, social division of labour has been division of work into occupations and not breaking up of the work activity into detailed operations. It may be noted that this sort of labour process prevailed even at the beginning of industrial capitalism.

Labour Process under Capitalist Relations

Labour process under capitalism is a unified process of creating use value and expansion of value (Nikitin, 1983). Under capitalist production labour process is directed towards the production of surplus value.

The capitalists purchase labour power for a wage and combine it with tools and raw materials and so on to produce commodities. In this process, surplus labour, embodied in the value of those commodities produced in excess of value of workers labour power expended in production, is generated. This labour surplus is realised by the capitalists by selling commodities on the market.

"The process of production, considered on the one hand as the unity of the labour process and the process of creating value, is production of commodities; considered on the other hand as the unity of the labour process and the process of producing surplus value, it is the capitalist process of production, or capitalist production of commodities" (Marx, 1978).

Appropriation of surplus value occur at two distinct phases of evolution of capitalist development at a formal and real subordination of labour to capital. The formal subordination of labour process to capital does not affect the form of production in the beginning. But the individual workers are brought together for co-operative production to enhance the productive potentiality which is realised by the capitalists through the exchange relations in the market. During this epoch, polarisation of capital occurs through the appropriation of absolute surplus value. Prolonging the working day and organizing production through simple cooperation, the size of the surplus value can be increased. This epoch forms the general ground work of the capitalist system, and the starting point for the second phase of real subsumption of labour.

Commodity production, circulation and exchange play an integral part in the production and appropriation of surplus value. Market is served by a number of different producers with commodities for sale. Competition forces producers to cajole to purchase commodities from one unit of production rather than from another. This induces each one to produce commodity for a price which is lower than other producers. This requires the producer to reduce the socially necessary labour time embodied in the commodity. The capitalist himself surrenders to these immanent law of capitalist production by revolutionising out and out the technical process of labour. In the course of this development, the formal subsumption is replaced by real subsumption of labour to capital[2].

The success of the capitalists to checkmate labour at this epoch gave them newer dreams. Why not control the work process too? After all labour is facile, its animation is uncertain. For the capitalists this was disquieting and undesirable. The increasing concern in this line snowballed into an important goal in capitalist scheme of things - the idea of wresting control of labour process. This urge emanates from the nature of labour power. Labour power is a commodity and the implied contract separates the labour from the labour power. Under the capitalist relations, the labourer loses his interest in labour power. However, what the capitalists buy is a potential to labour. In the apparently free contract entered into between the capitalist and the labourer, there is no express contract to ensure the full realization of the potential labour, i.e., labour power.

"What he buys is infinite in potential but in its realization it is limited by the subjective state of the workers, by their previous history, by the general social conditions under which they work as well as the particular condition of enterprise and by the technical setting of their labour" (Braverman, 1979).

It is in the interest of the capitalist to extract maximum labour out of it. Moreover, the reconstitution of labour process enable the capitalist to stymie any work resistance by the labourers. Thus the inherent contradiction between the dominant classes assumed newer dimensions. Capitalist development has reflected in its progressive epochs the conflicting relation in newer planes. The mutually contradictory interests of the dominant groups in the production process have made the labour process a 'contested terrain'. The inherent 'contested terrain' between the capitalists and the working class in the course of capitalist accumulation is portrayed by Edwards (Edwards, 1979). Edwards focused on the strategies and mechanisms which the capitalists have increasingly resorted towards the resistance, the labourers have formidably put against the capitalist attempts of controlling the labour process. Even though, Edwards have analysed these changes in the U.S.A's context, it was a fair description of changes under the capitalist system as a whole. His narration of history of about one and a half century of efforts and counter efforts of the two opposing classes shows the nature of the fight as continuous and inconclusive one. This has brought tremendous changes in the organisation and technological spheres[3].

An analysis of the labour process changes shows that the main mechanism the capitalist used to secure their goals to augment profit and capture work control was technology. Technology enters into capitalist scheme mainly at two levels. First, as an instrument to revolutionize capitalist production in order to maximise the difference between value of labour and value created by labour to augment profit conditions. Second, to embark technology in the work process in such a way as to embed work control mechanisms to establish capitalist control. The basic impulse of the capitalists to make epochs in technology thus stems from negative premises. Under capitalist relations technological progress is grounded upon a narrow, sectarian and lopsided perspective. However, the capitalists succeed in camouflaging their undesirable and negative approach towards technology by emphasizing the magnificent efficiency and productivity of the technology in creating the social product[4].

It is this negative connotation of capitalist technology that induced Braverman, while analysing capitalists attempts of control of labour process to remark, "that the problem can be fruitfully attacked only by way of concrete and historically specific analysis of technology and machinery on the one side and social relations on the other and of the manner in which these two come together in existing societies" (Braverman, 1979).

At a general level, the implication is that any study on issues of development must bring in its centre stage capitalist designs on the one hand and its fructifying technological upheavals on the other. Even though Braverman had in his mind a matured capitalist economy in this regard such a focus is essentially inevitable in analysing development issues in LDCs. The pertinent questions are:

1. what are the characteristics of the technology evolving in LDCs and under what type of social relations ?
2. is there any organic relation in the formation and growth between the two ? The socio-economic conditions spectacularly differ in developed and less developed economies. Leaving aside the details of the dynamics of segregation of the world economy into developed and less developed parts, it would be sufficient for our purpose to emphasise the Centre-Periphery relation that emerged in the history of capitalist development process.

Under this relation, the development perspective of LDCs will be greatly influenced by capitalist paradigms! Also, the structure and pattern of development efforts would be modelled after capitalist relations and through this capitalist contradictions and labour process changes are entrenched into LDCs. However, while the transition into different progressive modes of production occurred in developed capitalist economies linearly and gradually without causing disruptions in the organic relations and substance of such development, the change in LDCs occurred in a way that the latter economies were transformed into mere appendages to meet the needs and requirements of capitalist countries. They were destined to subserve the interest of the domineering capitalists and the latter in turn to obstruct or support a transition to higher forms suited to their interests.

Considering technology as the lever to clinche the capitalist designs, what are the ramifications it created in the LDCs under this historic relation? Also, what are its implications on the labour process in LDCs?

The intrusion of technology in LDCs under the Centre-Periphery relation occur mainly through technology transfers. The transmission of technology into LDCs mainly take place through the economic tie up emanating both from the activities of metropolis capitalist on the one hand and the domestic subservient capitalist on the other. While such technology transmissions assure easy profits for the capitalists, the benefits which LDCs generally accrue are nothing but large scale distortions and disequilibriums in its socio-economic milieau[5]. A fatal distortion which creates far ranging impulses in the economy of LDCs is the intensification of the dual character of the economy[6]. The vertical integration of capitalist technology into a particular sector creates a wedge in the form of a modern sector pushing back the original form of the sector as a traditional one incapable of any development prospects[7]. In that process, forces are released or formed which link the two separated parts in such a way that the modern one is poised for a perpetual growth at the cost of deterioration of the other. Also, in that process the benefits are appropriated by a few through marginalizing and depriving a majority who were linked organically with that sector. This in turn, causes further imbalances and distortions at other levels in the economy.

The effect of capitalist technology upon the labour process in LDCs also warrants close focus. The labour process changes in capitalist countries had occurred linearly in the realm of contested relation between the capitalist and the working class[8]. However, in LDCs such a pattern need not be replicated since there are structural distinctions in the contested relation between them. While the conflicting relation forms and persists in capitalist countries through the capitalist law of value, in LDCs, it is brewed from the interweaving of capitalist sector and traditional or pre-capitalist sector.

In other words, in LDCs, it is a conflict between two modes of production unlike the conflict of opposing classes under capitalism. The interaction or articulation[9] of modes of production turns into one of relations of contradiction and class struggle.

"If anything, 'articulation' specifies the nature of the contradiction. As Rey himself puts it, the idea is of the 'articulation of two modes of production, one of which establishes its domination over the other ... not as a static given, but as a process that is to say a combat between the two modes of production, with the confrontations and alliances which such a combat implies: confrontations and alliances essentially between the classes which these modes of production define" (Carter, 1978).

Apart from the structural differences at the modes of production level, the form of the contested relation is also different in capitalist economies and the third world. In the former, the basic issue of contest is about sharing of economic surplus. The capitalists system is increasingly accommodative of such demands since capitalism itself undergoes transformations which would enable them to find required resources elsewhere.

In the capitalist development process Marx has indicated three principal aspects of capitalist production (Marx, 1978).

1. The concentration of means of production into a few hands, making production more socialised.
2. The organization of labour itself as social labour, by co-operation, division of labour etc.
3. The creation of world market.

Further, Hilferding hinted at the development of capitalism as organised capitalism in which the forces of market cease to play a free role. He had distinguished four main features of such an economy first its basis in technological progress; second, the use of new opportunities in an organized way through cartels and trusts, third, the internationalization of capitalist industry and fourth, the replacement of free competition with scientific methods of planning (Bottomore and Goode, 1983).

The hints given by Marx and Hilferding about the course of development of capitalism had materialized as monopoly or state capitalism. Irrespective of the particular form, this transformation has succeeded in strengthening and intensifying the structure which the capitalist development had created in relation to LDCs. This development has facilitated in keeping LDCs effectively under subjugation. The regime of international trade and investment dominated by such capitalist development, enabled the capitalist economies draining away of resources from the LDCs as debt payments, transfer pricing, technology costs etc. While such transfers had gone a long way in protecting the interests of every section in the capitalist economy (in spite of conflicting relation among them), the same process had resulted in perpetual deprivation of the people of LDCs, particularly those on the fringes of the economy.

Under these different structural set up, it is unlikely that the unilinear changes as occurred in capitalist countries could take place in LDCs. Our concern then would be in locating the likely directions and patterns of changes. Whether the interaction between these modes would culminate in subsumption of pre-capitalist mode by the capitalist mode as suggested by the articulation theory implying passitivity of the pre-capitalist mode in the articulation process[10]?

It may be noted that there are writers who emphasise on the activity of the non-capitalist mode[11]. O' Laughlin writes "..... any general theory of imperialism can only be a theory of capitalism but any historical understanding of imperialism requires conceptualization of the dynamic of non-capitalist modes of production as well (Banerjee, 1985).

Considering non-capitalist mode of production as active in the dynamic process of articulation with the capitalist mode of production, we would say that the articulation process manifests dissolution and/or conservation of force; depending up on the degree of resilience each mode carries. Marx himself believed in the struggle between modes of production and pointed out that the outcome depends on their structures and specificities (Banerjee, 1985)

"People make their own history, but they do not make just as they please; they do not make it under circumstances chosen by themselves, but under circumstances directly found, given and transmitted from the past" (Marx, 1852). Contested relation under different modes of production will cause changes in labour process in unexpected directions. This is because first the new capitalist mode of production could not release forces whose impulses are capable of changing the entire sector since such forces are designed, regulated and controlled by alien capitalists. Second, capitalist intrusion makes only a partial and incomplete transformation of sectors making economic progress permanently retarded in the lagged ones. However, these incomplete transformations are incapable of wrecking the internal solidity and consistency of the evolved mode of production[12]. More than that the increased deprivation and marginalisation of people along with the retardation of capitalist development induce them to strengthen their production structure at a competitive level[13]. While the inner strength and consistency of an evolved mode of production provide stable development forces, the footing of this traditional mode at a competitive level with that of the new capitalist mode also induce them to receive and appropriate new forces of development that are consistent and conducive to the evolved mode. These peculiaries at the existence of multiple modes of production make the labour process changes in unbound and unknown directions compared to that of labour process changes in capitalist economies. It has been pointed out that in the absence of a winning mode the outcome of the articulation process depends on the specific factors involved in the concrete combination of modes of production in any particular society.

The labour process changes in traditional fisheries in Kerala give us indications of changes on these lines. A deprived and marginalised primary producers due to the capitalist intrusion, after their initial sufferings, began to draw inner strength particularly at production level. In this endeavour they also appropriated beneficial aspects of capitalist changes. These interactions of the native and the alien modes of production have provided power and strength to retrieve the labour process control of the traditional fishermen. The following chapters attempt to discuss these changes.

NOTES

1. Miomir Jasksic views that the theory of modes of production provide good basis for analysing issues of developing countries (Banerjee, 1985).
2. Real subordination of labour to capital implies the dominance of machinery in the labour process, incessant transformation of the labour process, and the imposition of strict factory discipline thus making the workers a 'living appendage' of the lifeless machine. Further, it is not the worker who employs the means of production but the machine employ the worker (Bottommore, 1985).
3. The capitalist development has brought into existence different control mechanisms like personal control, heirarchical control, technical control, and bureaucratic control consistent with the newer forms of capitalist development and the emerging forms of inherent class antagonism. The origin, form structure and the mechanisms of these controls are detailed in Chapter 2, 7 and 8 by Edwards (Edwards, 1979).
4. Capitalism is described in glowing terms for its productive powers. The bourgeoisie, during its rule of scarce one hundred years, has created more massive and more colossal productive

powers than have all preceding generations together (Communist Manifesto,). However given the negative impulses from which technology is emerging, the concepts of productivity and efficiency have to be put in social scrutiny. Issues like whose productivity and efficiency and who are the ultimate beneficiaries must be addressed in the whole social context.

5. The nature and determinants of capitalist technology and its inappropriateness in LDCs are discussed in Chapters 1 and 2 of Technology and Under development (Stewart, 1997).
6. Dual character is not dualism per se. Dualism implies two economic sectors, one advanced and other backward. However, both sectors operate independently of one another. They are in effect two separate economies. Dual character implies that the capitalist sector and the traditional sector (two modes of production) generate process of relations between them (Harod Wolpe, 1985)
7. Such relation/integration between capitalist sector and pre-capitalist sector has been termed in Marxian terminology as 'articulation of modes of production'. It was widely believed that in the articulation process the pre-capitalist mode of production would be subsumed ultimately by capitalist mode of production.
8. The unilinear changes in labour process in the realm of conflicting relation between capitalists and working class was analysed in the context of U.S. economy by Richard Edwards (Edwards, 1979). The different phases emerged in the capitalist technology and organisation of production in their attempts of control of labour process by capitalists entails changes exclusively in the capitalist mode of production culminating such changes as unilinear.
9. Articulation implies the combined presence of different capitalist and non capitalist modes of production.
10. Major theories of development neglect the primacy of the original modes of production, in favour of universal stages, visualizing former ones as passive.
11. There are three models which depict the likely changes of articulation process in the context of capitalist development. One possibility is that the various modes of production

simultaneously exist but principally independent of one another. The second model is that various modes of production in a society are interrelated under the dominance of one of these modes of production (probably capitalists). The third model holds that the modes of production are interacting in such a way that there will be no dominant mode of production (Wilber, K.C; and Jameson, K.P; 1992).

12. Marx also spoke about the internal consistency and solidity of pre-capitalist modes of productions in Asian countries such as India and China.
13. Marx had postulated a struggle between the expansive urges of capitalist mode of production and conserving forces of pre-capitalist modes of production, i.e., struggle between diametrically opposed economic systems (Banerjee, 1985).

REFERENCES

1. Banerjee, Diptendra, (1985): *Marxian Theory and the Third World* Sage Publications, New Delhi.
2. Bottomore, T.B., (1990): *Theories of Modern Capitalism*, Universal Book Stall, New Delhi.
3. Bottommore, Tom, and Patrick, Goode, (eds.), *Readings in Marxist Sociology*, (1983): Clarendon Press, Oxford.
4. Braverman, Harry., (1979): *Labour and Monopoly Capital, The Degradation of Work in the Twentieth Century*, Social Scientist Press (Indian Edition), Trivandrum.
5. Carter, A.F., (1978): "The Modes of Production Controvercy" *New Left Review, 107.*
6. Edwards, Richards., (1979): *Contested Terrain, The Transformation of the Work Place in the Twentieth Century*, Basic Books, Inc., Publishers, New York.
7. Marx, Karl and Engels., (1848): *Communist Manifesto. Friedrich,*
8. Marx, Karl., (1852): "The Eighteen Brumaire of Louis Bonaparte", *in the Marx – Engels Reader, Robert C. Tucker, New York. (First published in 1852)*
9. Marx, Karl., (1978): *Capital, A Critique of Political Economy*, Volume I, Progress Publishers, Moscow.

10. Miles, Robert., (1987): *Capitalism and Unfree Labour Anomaly or Necessity ?* Tavistock Publications Ltd., London.
11. Nikitin, P.L., (1983): *The Fundamentals of Political Economy,* Progress Publishers, Moscow.
12. Stewart, Frances., (1977): *Technology and Underdevelopment,* The Macmillan Press Ltd, London.
13. Volkov, M.L., (1985): *A Dictionary of Political Economy,* Progress Publishers, Moscow.
14. Wilber, K.C., and Jameson., (1992): *The Political Economy of Development* K.P., (1992): *and Underdevelopment,* Mc.Graw - Hill, Singapore.
15. Wolpe, Harold., (1985): "The Articulation of Modes and Forms of Production", *in Marxian Theory and the Third World. Diptendra Banerjee, (ed.), Sage Publications, New Delhi.*

3

Fishing Technology : A Back Drop

In this chapter, we endeavour to present a brief summary of technical changes of primary production in the marine sector. This enable us to understand the physical changes in chronological order and the type and degree of such changes. A physical description of the instruments of production is essential as it throws light on the capital requirement which illuminates the social economy of the fishery sector. Further, such an analysis helps us in identifying the forces instrumental in bringing the changes.

In the fish harvesting the major means of production involved are:

1. crafts,
2. gears and its accessories, and
3. methods of fishing.

Depending on the surf conditions, nature and availability of fish stock and the relative economic conditions of fishermen, the use of the instruments of production vary in coastal regions.

In coastal Kerala, the major types of crafts used in traditional fishery are :

1. catamarams,
2. dugout canoes and,
3. plank built boats.

Catamarams are mainly used by fishermen residing south of Kollam stretching up to coastal areas of Kanyakumari district in Tamil Nadu. Dugout and plank built canoes are used by fishermen all over coastal Kerala. However, their predominance vary in different districts by the influence of specific local conditions there.

Catamarams

Catamarams are traditional crafts used by fishermen of eastern coastal sides of India from Orissa to Kanyakumari. Later, its use was extended to Thiruvananthapuram and Kollam. Catamaram, a keel less craft is formed by tying together few logs of lightwood with coir ropes Two wooden supporters called catamarams are used for lashing them together (Mathur, 1978). Kadamarams are of two types, of 2 feet and 1.4 feet respectively. When the logs are tied together they become curved and shaped like a canoe. Catamarams are broadly of two types, the big one 7.50 to 8.50 metres long and 0.80 metres wide and the small one 4-5 metres long and 0.60 metres wide (Korakandy, 1994). The former accommodates three to four fishermen as a unit and the latter is operated by one or two fishermen. The investment required for a new catamaram is Rs. 10000 - 16000 varying upon its size[1]. A technical speciality of a catamaram is that it is a versatile craft and can be used almost on all seasons at all points on the shore.

Dug-out Canoes

The dug-out canoes are made by scooping out the wood from a single log of mango or jungle jack. The keel portion is thicker than the sides. It is shaped by using teak panals if necessary. The dug-out canoes may be large or small. The large ones are 9.50 - 12.50 metres long, 0.90 - 1.50 metres wide and 0.75 - 0.90 metres deep. The small ones are 7.20 - 8.50 metres long 0.90 -1.20 metres wide and 0.45 - 0.60 metres deep. Seven to eleven fishermen can work on large dug-out canoe where as the smaller one can accommodate three to six fishermen. The investment requirement on bigger dug-out canoes are high. A bigger dug out canoe now costs Rs.30000-60000 depending up on the material used. The dug-out canoes are used eight months in a year from October to May (Directorate of Fisheries, 1969). Dug-out canoes are modified into high board dug-out by adding planks stitched to the sides of the dug-outs.

Plank-Built Boats

Plank-built canoes are constructed by seaming together planks of wood using coir ropes and copper nails (Bhushan,1979). They are made with or without ribs on the sides. Black pitch coating is used to make them water tight. These undecked crafts are also found in two sizes. The large ones measure 11-13 metres in length 1-1.5 metres in width and 0.70-0.80 metres in depth (Korakandy, 1994). The small ones are 6 - 9.50 metres in length, about 0.90 meters in width and about 0.68 meters in depth. The large ones are operated by 12-15 fishermen while the small ones carry a crew of 4-6 persons. The large ones are used from July to October. The smaller ones are used from September to March/April. The various characteristics of the crafts used in the traditional sector is summarised in table 3.1.

Gears

In the traditional sector, numerous fishing gears are used by fishermen. The technological flexibility of traditional fishermen in the fish tackling is learned from the diversity of crafts and gears used in the fishing process. The gears have been evolved from the knowledge gained from the long experience of the shooting and feeding habits of each specie of fish stock. These gears are used in different combinations with the crafts depending up on the seasons, availability of fish, and biological characteristics of species.

In the case of nets, generally it is the mesh size, the thickness of the yarn with which the net is fabricated and the shape of the net that influence the nature and size of the fish caught in it. In the case of hook and lines, the size of the hook and the length of the line are the important determinants.

A general description of the gears used is made to show the flexibility of the technology and the ichythyological knowledge of the traditional fishermen. There are basically three types of nets used in the fish tackling process besides different types of hook and line sets. These are:

1. shore seines,
2. boat seines and
3. gillnets.

Shore Seines

The Shore seines are bag shaped nets operated from the shore with the help of a canoe. The working of the shore seine is such that one wing of the net remains on the beach, and the other wing are taken out in the canoe, drawing it in a semi circular manner and finally bring the other end to the shore. After the net has been laid, the two ends are simultaneously and gradually pulled in by fishermen (Kurien, 1978). A canoe with a crew of six to eight persons are used to place the net in the sea and twenty five to forty persons are employed for pulling the net. Shore seines are used all along the coast of Kerala. It is used about six months during calmer seasons between November and March/April. It is the pelagic and shoaling fishes that are caught with it. (Korakandy, 1994). The use of shore seines are fast failing in coastal Kerala at present.

Table 3 :1 : Major Characteristics of the Crafts in the Traditional Sector in Kerala.

Crafts and size	*Cost per unit (in Rupees) (at 1996 -1997 prices)*	*Life span (Years)*	*Crew Size*	*Gear used*	*Seasons of use*	*Area of use (Kilometeres)*	*Distance fished*	*Depth range*
Catamaram (Large) Length 7.50-8.50 (meters) Width 0.80	10000-16000	10	3-4	Gillnets and hook and lines	Any season of the year	South Kerala	12-15	Upto midlayer water of the sea
Catamaram (Small) Length 4-5(meters) Width 0.60	5000-8000	10	1-2	Gillnets and hook and lines	Any season of the year	South Kerala	12-15	Upto midlayer water of the sea
Dug out Canoe (Large) Length 9.50-12.50 (meters) Width 0.90-1.50 Depth 0.75-7.90	30000-60000	20-25	10-15	Boat seines and shore seines	October - May	All Kerala	12-15	Upto Midlayer water of the sea
Dug out Canoe(Small) Length 7.20 - 8.50 (Meters) Width 0.90-1.20 Depth 0.45-060	15000-25000	20-25	3-6	Gillnets and hook and lines	October - May	All Kerala	12-15	Upto Midlayer water of the sea
Plank Canoe (Large) Length 11-12 (Meters) Width 1.00-1.50 Depth 0.80	20000-60000	10-15	12-15	Boat seines Gillnets. shore seines. hook & lines	July - October	All Kerala	12-15	Upto Midlayer water of the sea
Plank Canoe (Small) Length 6.00-9.50 (meters) Width 0.90	15000-25000	10-15	4-6	Boat seines. gillents. shore seines. hook and lines	September-April	All	12-15 Kerala	Upto Midlayer water of the sea

Source : (1) Bhushan, (1979), (2) Korakandy. R., (1994), (3) Mathur, (1978)

Boat Seines

Boat seines are a kind of encircling nets (Sainsbury, 1971). They are conical, bell shaped or bag-shaped nets made from cotton, hemp or nylon. The open end of the boat seines normally have larger meshes which decreases in size towards the closed end. It is operated using two boats, canoes or catamarams which pull at the ends of the two wings of the net. This keep the mouth of the net open and the fish swim into the narrower end. In this process scaring devices made of coconut leaves or wood are often used to produce sound in the water which drive fish into the mouth of the net. Depending upon the size of the net and the craft used, a boat seine can be operated by as few as five persons to as many as twenty persons. Boat seines are used all over Kerala and normally used to fish pelagic and mid water shoaling species. It is generally shot at a depth of 10 to 20 fathoms.

Gillnets

Gillnets are single walled nets and are of different types. Gillnets are of set, floating or drifting types depending up on the way they are used (Kurien, 1978). Set gill nets are used from stationery crafts and can be set either at the surface or at the bottom. Floating gill nets are suspended in water with anchors at the bottom and floats on the top. When a floating gill net tied to a craft is allowed to drift with ocean current, it is known as drift net (Bhushan, 1979). Fish is get caught in the gillnet when they swim in to it and gills get entangled in the mesh of the net. The gill nets with different mesh sizes catch different specie like mackeral, seers, eel, cat fish, skates and ray, sharks etc.[2] While gillnets are used all along Kerala coast, they can be operated with as little as two persons as a catamaram unit or as many as twelve as a canoe unit depending up on the length and weight of the net.

Besides these nets, some other type of nets such as stake nets, Chinese nets and cast nets are also used in fishing.

Hooks and Lines

The hook and line fishing is the most commonly known method of fishing. The Kerala fishermen have been used this method of fishing from time immemorial (Bhushan, 1979). This method is used generally for fishing in deeper waters to catch mainly sharks, eels and seer fish. The type of the fish caught depends up on the depth to which the line is sunk and the size of the hooks.

Three types of fishing lines are used by Kerala fishermen; hand lines, long lines and chain lines[3]. The hand line represents the simplest method of fishing and are generally cast from anchored canoes in shallow as well as deep waters of the sea(Kurien, 1978). The long line consists of a master line with equi-distant thinner branch lines to which the fishing hooks are attached. The number of hooks attached depends upon the length of the line. The chain lines are used for catching sharks and they use specially strong and hook lines (Mathur, 1979). Some of the main characteristics of the gears used in the traditional sector of the marine fishery are described in the following table.

Table 3.2. Major Characteristics of Gears in Traditional Marine Fishery

No. Gear	*Average size (length in metres)*	*Mesh size (in cms)*
1. Fixed nets	12-30	1 to 2 at Cod end.
2. Stake nets		
Seine nets		
(i) Boat seines		
(a) Kolluvala	73	1 at cod end 2 at mouth
(b) Thanguvala	50-65	2 at cod end
(c) Madivala	49	2 at cod end
(ii) Shore seines		
(a) Kambavala	316	0.80 at cod end
(b) Aray nets	3.60-18.30	0.60-1.20
3. Cast nets	2.50-6 in radius	1.20
4. Drift nets	48-125	5-6
5. Long line and hand line	Several hooks are used depending in the length of the line	-

Source: Bhushan. (1979).

Technology in the Modern Sector

Fishing techniques used in the modern fishery are very complex. The multinational companies involved in seafood business operate 'floating factories' in ocean waters using computer and laser technology. However, their operations are limited by the maritime sovereignty of different nations even though such multinationals are able to scuttle such restrictions through their clout under the "globalisation fever' in which LDCs are trapped. Thus we see in the fishery sector the existence of century old simple technique of the use of catamaram in one extreme to the operation of "floating factories' on the other. But it may be noted that our concern is to focus on the modern technology existing in the fish economy of Kerala. Even though, globalisation policies of Govt. of India have resulted in the proliferation of multinationals in the deep sea fishing in Indian waters such technical configurations have not pervaded into the fish economy of Kerala. More over, the Kerala fisheries has its own history of modernisation attempts.

The modernisation programme in Kerala fisheries started in 1953. Mechanization process in the capture fisheries confined basically at three levels.

1. Craft movements (method of propulsion)
2. Development of gears and
3. Tackling techniques.

These technical changes considerably improved the productive capacity of the fishing sector. The new techniques raised the productive capacity specifically at four levels. First, the use of machine power enabled the fishermen to reach the fishing ground early and this has raised the fishing time. Secondly, the new technology enabled the fishermen to increase the distance range of fishing operations. Thirdly, they succeeded in capturing the bottom dwelling or crustacean species like prawns, crabs, lobsters etc. as the new technology has raised the depth range of operations. Fourthly, the lessening of the fatigue element of the work character also contributed to raise productivity.

The mechanization efforts in the Kerala fishery is divided into two distinct periods:

1. 1953 -1963 and,
2. the period there after.

The distinctness stems from the fact that the former attempt was made under the guidance of foreign assistance whereas the latter was purely an indigenous attempt. The first phase of the mechanisation effort was limited to the mechanisation of crafts. This effort was carried out with the help of foreign assistance involving various agencies mainly under the Indo - Norwegian Project (INP) and experts and naval architects associated with Food and Agricultural Organisation (FAO). The attempts under these agencies lasted for a decade until 1963, when the FAO experts submitted their last and final report on the mechanisation of fishing boats in India. However, it changed towards exploratory and experimental fishing. Since 1963, as pointed out earlier, the mechanisation efforts were largely indigenous.

A description of these modernisation efforts, the individuals and agencies associated with it are required since the results of their efforts had created wide disruptions especially on the social fabric of the fishery economy. We would be able to see a class polarisation as the main outcome of these efforts. Hence an investigation is made on these efforts in detail.

Agencies and Institutions Associated with Mechanisation Efforts in Kerala; The FAO/EPTA Attempts

The unsuitability of foreign technology in marine fisheries in Indian and particularly in Kerala conditions need to be learned from the experiences and efforts of the foreign agencies in India. The Government of India appointed a committee consisting of two experts from FAO/EPTA[4] - Paul.B.Ziener and Kjeld Rasmussen to advise on improvements to existing boats with regard to design, construction, safety rules and engineering. They were also required to advise on the mechanisation of available boats and to design new improved type of fishing boats.

These experts undertook a survey of all traditional crafts of the maritime states in India. They also examined various types of engines that could be used to motorise traditional crafts. However, their attempts succeeded only partially. This was especially true in the case of crafts operated along Kerala coast as they were found either too small and narrow or lacked stability. The attempts at modification of hull design also failed. It was admitted that it was most difficult and impossible to mechanise catamarams and canoes (FAO, First Reports 1958). There was a general feeling that the "logical step to take" would be the introduction of mechanised beach landing crafts or what were called the surf fishing boats to replace catamarams and canoes (FAO, First Report, 1958). The FAO experts reached the conclusion that the only possibility of carrying out mechanised fishing from long surf beats coasts seems to be in the development of surf boats.

Following this, in 1952, FAO employed a special consultant, a Norwegian naval architect, Hans.K.Zimmer and he found that none of the surf boats in Europe suited to Indian conditions and hence was decided to develop designs of mechanised surf boats suitable to Indian conditions (FAO, First Report, 1958).

Between 1954-58, the FAO experts in India tried three prototypes of mechanised surf boats. But each one of these prototypes developed technical snags and above all operations from these prototypes proved to be financially unviable.

In 1958, the FAO experts in their report to Government of India commented that it is impossible to release any final design of a surf boat for Indian conditions as much work remains to be done before an economical and practical size of boat is developed. Followed by this report, in 1959, the FAO, the Government of India through Central Institute of Fisheries Technology (CIFT), Cochin and INP, resolved to pool their resources in trying to design a suitable mechanised surf boat. Four different types of surf boats were built and tested under this arrangement till 1963[5]. However, after 1963 no work has been done on surf boats. Thus despite a decade of efforts in the development of a viable surf boat, there was none in use particularly along the Kerala Coast[6].

Mechanisation Efforts under INP

The INP started functioning in Kerala, was formed out of a tripartite agreement for economic co-operation between Government of India, the Government of Norway and the United Nations in October (Master Plan, 1969). A supplementary agreement was signed in 1953 for establishment of a project for development of fisheries. One of the specific objectives of the project was making improvements in methods of fishing[7]. In this regard INP also first tried to.make changes in crafts. The project focused its operations in two fishing villages -Sakthikulangara and Neendakara in Kollam district. In these two villages, the project never tried to motorise or mechanise the traditional crafts (Master Plan, 1969). Instead boat designs were imported from Norway and a series of different sizes of mechanised boats were constructed at Neendakara. The project evolved a number of small mechanised crafts like 22ft; 23 l/2ft, 25ft and 28ft and issued to local fishermen. The following table shows the various types of boats that were issued to fishermen under INP project.

Table 3.3. Mechanised Boats Introduced by INP (1953-1963)

Length of the boat	*Horsepower*
22ft	8
23.5ft	8
25ft	8-16
30ft	36
36ft	48

Source: Bhushan, (1979)

Another technical improvement made by INP was the introduction of trawlers designed by INP and powered by a 48 H.P. engine which is of 36 ft (10.8 metre) stern trawler. This was exclusively meant for shrimp trawling. The development of such a boat was the outcome of expansion of export potential for shrimps.

Development of Indigenous Technology

From 1963 onwards, the technical changes in the fishing sector was indigenous in nature[8]. The technical changes during this period were in response to the changes in the economic sphere of the marine fishery sector. The development of export market[9] especially for shrimps have resulted in large volumes of capital flowing into capture fisheries. The need for fishing vessels increased considerably. This increased demand has resulted in the construction of medium and large sized mechanised boats. The local capitalists developed indigenous engines for mechanised boats. The legacy of Research and Developments (R & D) carried out by foreign institutions and experts was continued by national institutions with in the "technological standards" determined by them[10]. After 1963, most of the research was carried out under the auspices of the CIFT, Cochin.

The response at institutional level to the changing economic environment is reflected in the standardisation of new mechanised boats by CIFT. Between 1963 and 1967, the CIFT has standardised four new mechanised boats:

1. a fishing boat,
2. a trawler,
3. a drifter/trawler, and
4. a combination vessel which could be used for seasonal trawling as well as other kinds of fishing.

It may be noted that the parameters of development of mechanised boats were the same as that fixed by INP/FAO set up. No attempt was made by CIFT to equip the existing traditional crafts with mechanised propulsion system or other ancilliary developments. The CIFT 'progressively' believed in the INP/FAO experts advise that the traditional crafts were unsuited to mechanisation.

The technical characteristics of the standard types developed by CIFT were such that it could fulfill the commercial motives. The first one, the fishing boat had a length of 30 ft and was fitted with a 30-35 h.p. diesel engine. Its depth range of operations was set up to 15 fathoms; it had a crew requirement of six persons and it could stay at sea for 20 to 24 hours. The other three varying from a 40 ft trawler to 50 ft combination vessel using 80-90 h.p. and to 150-160 h.p. diesel engines and were able to stay at sea from three days to a week and capable of fishing up to much greater depth ranges of 25-30 fathoms.

We may also look into the capital costs required per unit of the mechanised boats standardised at CIFT.

Table 3.4. Costs Per Unit of the Mechanised Boats Standardised at CIFT (at 1977 price level)

Boat Size	*Type of Boat*	*Cost/Unit (Rupees)*
25ft	Open Fishing boat	37,400
30ft	Fishing boat	68,700-76,700
32ft	Trawler	86,000-1,04,200
32ft	Fishing boat	86,000-1,04,200
36ft	Trawler	1,45,000-1,53,000
40ft	Trawler	1,64,000
45ft	Drifter/Trawler	1,96,000
50ft	Combination vessel	3,40,000

Source: Bhushan (1979).

In the development of these medium size boats, a point to be noted is that, the latter types were capable of both trawling and other types of fishing ie, boats were capable of multiple methods of fishing. The machine manufacturers in India also came forward to make use of the commercial opportunity emerged in the marine fisheries. As we have noted earlier, all technologies required for modernisation in the fisheries sector were imported before 1963 and since then indigenisation occurred but with in the frontiers contoured by foreigners.

Changes in Fishing Methods

The technical changes were extended not in the development of crafts alone. New methods of fishing were also introduced. Some of the fishing methods developed along with the improvements in crafts were :

1. gill netting
2. boat seining
3. bottom trawling
4. pelagic trawling or purse - seining
5. long lining
6. lift netting
7. pumps fishing

There is no substantial difference in the process of fishing with gillnet and boat seines in mechanised boats with that of traditional fishing except for the fact that in the former, the sizes of nets will be bigger. How fish are caught with gillnets can be learned from figure 3.1.

Fig . 3.1. Gillnet Fishing

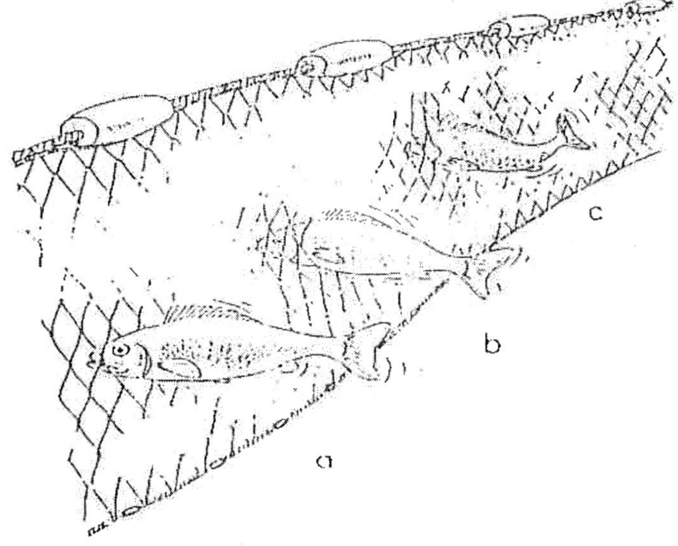

Source : FAO, (1980).

Bottom trawling is an active fishing technique done for prawn harvest. If a trawl net is pulled along behind a boat in the sea bed, it is trawling (FAO, 1980). A trawl net is a large one having a bag at the end of the net. It is wider at the open part and tapering through the body of the net to the closed end. The fish are trapped at the closed end. The mouth of the net looks like an oval opening when viewed from the front, and the two wings of the net stretch out in front on either side to widen the area swept. The floats are fixed around the upper edge of the mouth along the headline. Around the bottom of the mouth is the ground rope which is weighted to remain at the bottom. Horizontal spread of the mouth of the net is attained by the 'otter boards' or doors towed ahead of the net and set at an angle of attack to the towing direction, thereby providing the outward force necessary to spread the wings to which they are fastened.

Trawling operations are organised in different ways, each one giving specific advantages. Mainly, the bottom trawling is categorised in to:

1. stern trawling,
2. otter trawling,
3. out - rigger trawling and
4. pair or bull trawling.

In the otter trawling, a large trawl net whose sides are held open by otter boards is capable of fishing more because of its flexibility to side ways. Since other trawling require huge engine power and hence it is not popular in Kerala coast. In the stern trawling a single trawl net is towed on the sea bed from the stern of the vessel. In stern trawling the craft is maintained on a straight course while hauling and setting and the pull is along the direction of the motion of the craft. The specific advantage of this trawling is that some of the voyage time can be used for fishing thus lessening fuel costs.

Fig 3.2. Stern Trawling

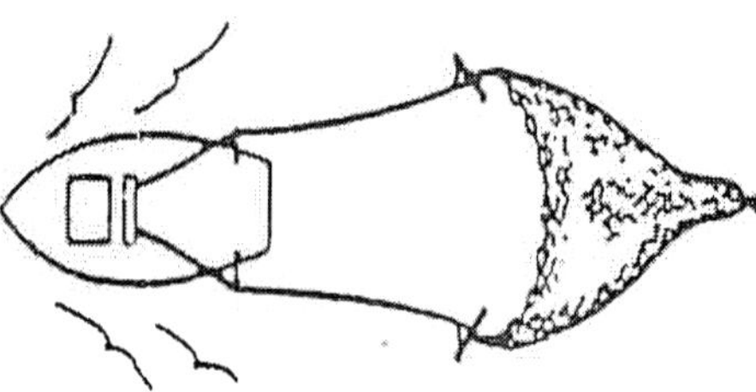

Source: FAO,(1985)

Figure 3.2. shows the stern trawling. In outrigger trawling, the outrigger booms are tied to the main mast of the vessel with trawl nets on both sides. The nets are towed from the ends of the outrigger booms on each side of the craft. Under this type of trawling, using the power required for a single trawl net, two trawl nets can be used. Figure 3.3. shows the outrigger trawling.

Fig. 3.3 Out -rigger Trawling

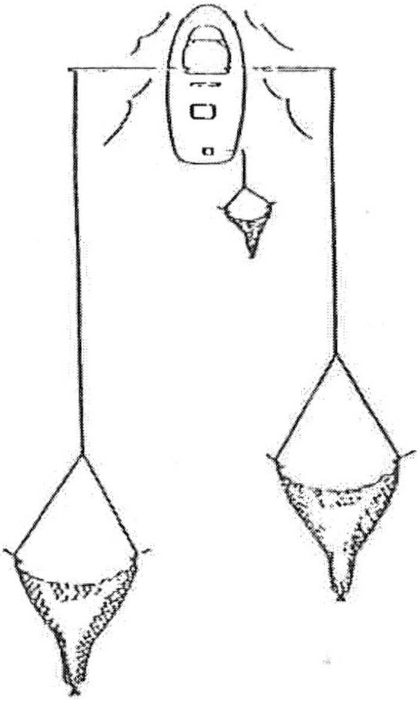

Source: FAO(1985)

In bull trawling two boats pull the trawl. The mouth of the net is kept open by the outward pull provided by the correct lateral spacing of the vessels. This method has the advantage of using a large net and also can catch more fish. This is because a single boat towing in front and at the centre of a trawl net may frighten some of the fish away with the noise of its engine while two boats towing in front and at the sides of the net will be making noises which will scare the fish towards the centre and straight in to the net. It is illustrated in figure 3.4.

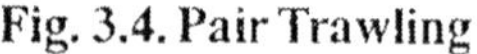

Fig. 3.4. Pair Trawling

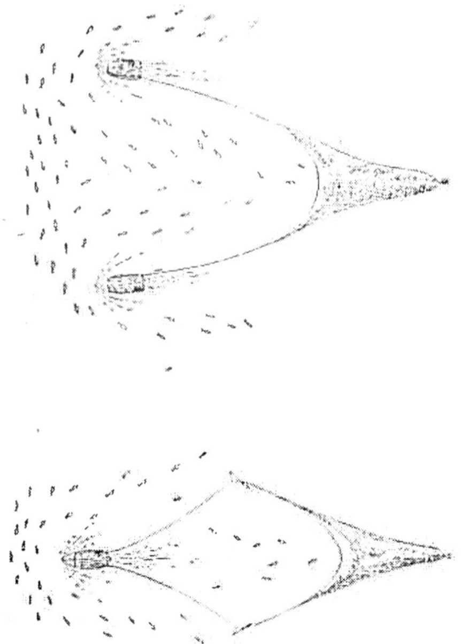

Source : FAO. (1980)

Peleagic or mid water trawling is used to catch peleagic species found in the mid waters. For catching peleagic fish, purse seine technique is used. It involves the setting out of a long net to form a meshed wall around the shoal of the fish which has been spotted. When the net has encircled the fish, its bottom is pulled together to hold the catch. The floatation is provided by a large number of floats fastened to the float-line. Weights are fixed on the lead line which runs along the bottom of the net to sink the net so that it forms the desired wall. Below the lead line, a purse line runs through rings connected by short length of rope to the lead line. The purse line is pulled from the pursing winches through the rings in order to close up the bottom of the net. Figure 3.5 shows the purse-seining. A severe defect of this method is that since the mesh size of the nets are small, it affects indiscriminate fishing of even small and spawning fish. Long lining fishing techniques has different varieties. In Kerala, it is hand lining which is more popular.

Fig. 3.5. Purse-Seining

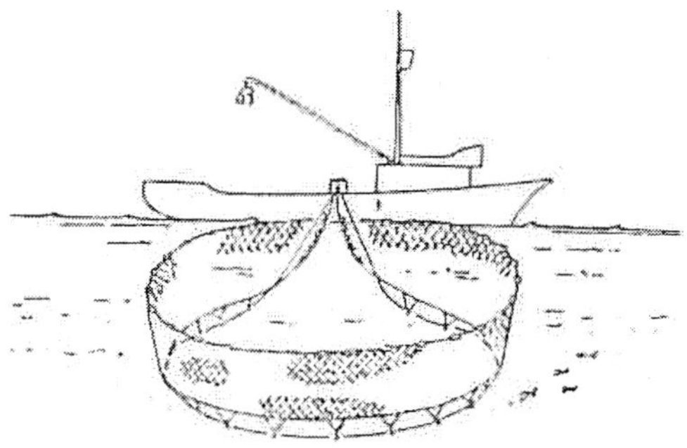

Source : FAO, (1980)

It may be noted that when nature of crafts and fishing methods have undergone changes, such changes affected the making of gears also. The use of new materials in the making of gears enhanced the durability and given enormous flexibility in the mesh size of nets. Thus we see a total change in the form of a 'technical package' in the modern fisheries covering the crafts, gear and new methods of fishing.

Development of Intermediate Technology

A discussion about technical changes in the fishing sector of Kerala will remain incomplete unless we look in to the development of 'intermediate forms of technology'. Technical changes at this level mainly related with improvements both in crafts and gears. In the sphere of craft improvements we see the introduction and massive spread of plywood boats and use of outboard motors both in plywood boats and adapted traditional canoes. Quite similar to that of mechanised sector, certain trawl and purse seine nets were also developed in this sector.

Thus in the fish economy as a whole, we see the prevalence of an amalgam of different techniques ranging from traditional to modern and an intermediate technology in between. The formulation and concretization of these different epochs of technology is beset with complex societial implications.

In the fishery sector of Kerala, modernisation attempts were done before the introduction of planned development by Travancore-Cochin governments. The main focus of such attempts were to increase the productivity of the fishermen and ameliorate their economic conditions. These erstwhile governments supplied cheap wood for canoe constructions, provided quality materials for gear making and helps were given to improve their marketing capacity (Hakkim, 1980). The official approach to fisheries development was centered around the fishermen community. However, this approach was later undermined by the modernisation programme. A set of technical equations were evolved with a common character both under foreign and domestic influences to intrude in this sector. Thus a technologically neutral approach of modernisation is difficult without uprooting a 'traditionally evolved equilibrium. In other words, technical changes have a class character and this aspect will become clear more when we examine the different phases of development experience in Kerala fishery. All these changes point towards the necessity of a theoretical design capable of explaining theses changes more scientifically. Braverman has pointed out that the study of technological change can be done from two sides, viz. from an engineering and a societial perspective (Braverman, 1974). In an underdeveloped economy it is the latter approach that must be followed particularly when we know that the technical changes are not neutral. In the next chapter we take up these issues in detail from the perspective of labour process changes.

NOTES

1. Information provided by fishermen during 1996-97.
2. The gillnets, to catch specific species are known locally in different names. The gillnet used to catch mackerel is known as Ayila - Calavala, for seers, eel and catfish as Olukkuvala, for skates and rays as Thirandivala, for white baits as Nettolivala, for sharks as Sravuvala (Korakandy, 1994). For local names of different types of gears used see SIFFS study, 1991.
3. The local names of the handlines, long lines and chain lines are Kaichunda, Beppu or Ayiram chunda and Changala chunda, respectively (Korakandy, 1994).
4. Under resolution 222(IX) A of the Economic and Social Council of the United Nations, the Food and Agricultural Organisation (FAO) gave technical services under its Expanded Programme of Technical Assistance (EPTA).
5. Details of the surf-boats constructed between 1959 and 1963 is available in the Third Report to the Government of India on Fishing Boats (FAO, Rome, 1963).

6. The surf-boats developed were economically non-viable has been affirmed by the First Report to the Government of India on Fishing boats based on the works of navel architects Paul.B.Ziener and Kjeld Rasmussen (FAO, 1958).
7. Besides this objective, the project was aimed at:
   i. increasing the profit of producers by better utilisation of produce, and
   ii. Improving the health and sanitary conditions of population and bettering standard of living of the fishermen community.
8. Major focus of research and development activity during this period of indigenous thrust was mainly on:
   (i) designing new mechanised crafts,
   (ii) development of indigenous engines,
   (iii) searching alternative material for boat building
   (iv) finding new materials and designs of nets and,
   (v) new methods of fishing.
9. During the period between 1964 to 1979, the quantity and value of shrimp exported from the country (mainly from Kerala) had increased from 8007 tons to 53669 tons and Rs.38.10 million to Rs.2237.92 million respectively. The change amounted to an increase of 6.7 times in terms of quantity and 58.7 times in terms of value (Korakandy, 1994).
10. For example, a Craft and Gear Division was organised in the Central Institute of Fishery Technology (CIFT), Cochin on the recommendation of the FAO experts, which conducted most of the research and development efforts in this area (FAO, Third Report to the Government of India, 1963).

REFERENCES

1. Bhushan, Bharat, (1979): Technological Change in Fishing in Kerala -1953-1977 (Unpublished M.Phil Dissertation) Centre for Development Studies, Trivandrum
2. Biswas, K.P., (1990): A Text Book of Fish, Fisheries and Technology, Narendra Publishing House, Delhi.

3. Braverman, Harry (1974): Labour and Monopoly Capitalism: The Degradation of work in the Twentieth Century, Monthly Review Press. New York.
4. Directorate of Fisheries, (1969): Master Plan for Fisheries Development, Kerala State Fisheries Directorate, Trivandrum.
5. FAO, (1958): First Report to the Govt. of India on Fishing Boats based on the Works of Kjeld Rasmussen and Paul, B-2einer, FAO, Rome.
6. FAO, (1963): Third Report to the Govt. of India on Fishing Boats based on the Work of Peter Gunner, FAO, Rome, 1963.
7. FAO, (1980): Fishing with Bottom Gillnets, FAO, Rome.
8. FAO, (1985): Definition and Classification of Fishery Vessel Types, Fisheries Technical Paper, 267, FAO, Rome.
9. Hakkim, A.V.M., (1980): Mechanisation and Co-operative Organisation. Their Impact on Traditional Fishermen of Kerala (Unpublished M.Phil Dissertation) Centre for Development Studies, Trivandrum.
10. Korakandy, Ramakrishnan., (1994): Technological change and the Development of Marine Fishing Industry in India, Daya Publishing House, Delhi.
11. Kurien, John, (1978): Towards an Understanding of the Fish Economy of Kerala State, (Working Paper No.68), Centre for Development Studies, Trivandrum
12. Mathur, P.R.G., (1978): The Mappila Fisherfolk of Kerala, Kerala Historical Society, Trivandrum.
13. Sainsbury, S.J., (1971): Commercial Fishing Methods. An Introduction to Vessels and Gears, Fishing New (Books) Ltd, London.

4

Labour Process Changes in the Marine Fishing Sector

The production process in the traditional sector involves organisation of production by fishing units as single independent entities. It resembles a competitive firm in modern manufacturing. A fishing unit consists of a team of fishermen with their craft, nets and other accessories. The number of fishermen in a unit varies depending upon the method of fishing used. Fishing process in the traditional sector is a skilled and complex operation where the workers participate at different levels in different capacities. The economic status of the participants in terms of ownership in the fishing process are:

1. the owner non-worker;
2. the owner-worker; and
3. the workers who are generally in the status of hired labour.

The size of investment requirements in fishing equipments is the main determinant of the relative position of fishermen in the participation of fishing operations. It is also a main factor reckoned in sharing the net out put. Generally the nature of ownership in the fishing sector is classified into two:

1. individual Ownership and
2. collective Ownership

Normally individual ownership predominates if investment requirement is low; otherwise collective ownership. Under individual ownership the participants in the fishing process are mainly the owner workers and the non-owner partners or workers. Under collective ownership the participants are the owner workers and non-owner labourers. Lenin and Mao Ze Dong have developed a scheme for analysing the rural classes and the latent class conflicts in the traditional agricultural sector on the basis of their accessibility to means of production (Bagchi, 1984). These classes are:

1. the land lords
2. rich peasants
3. middle peasants
4. poor peasants
5. proletariates

Following these pattern, in the traditional marine fisheries, on the basis of their accessibility to means of production, the fishermen can be grouped as (Klausen, 1968);

1. absentee fishermen
2. rich fishermen
3. middle fishermen
4. proletariates.

The absentee fishermen are the owners of fishing equipments who do not panicipate in any of the fishing activity except in organising the production process in fishing sector. They earn their income as rents on their equipments, commissions and interests and other types of payments for the services discharged.

The rich fishermen are those who own big crafts and gear and employ wage labour besides their own labour to conduct fishing activity. The middle fishermen own and operate crafts and gear which require lower investments. The proletariats receive their income by selling their labour power alone.

The organization of production in the traditional sector is motivated to earn a subsistence. Emergence of an absolute surplus is not a consequence of pre-designed commercialization. However, there are customarily determined appropriation mechanisms of these surpluses. Drawing on these surpluses, accumulation occurs on a limited scale. This accumulation process doesn't create any wide cleavage in the economic life of the fishermen. The main reason for an almost egalitarian economic structure is on account of low investment requirements in certain type of fishing equipments making them accessible to the subsistence sector. The observed pattern of ownership existing in the traditional sector is shown in the descending order of the egalitarian relationship between the owners and non-owners in Table 4.1

Table 4.1: Ownership Pattern in the Traditional Sector

1	Individual Ownership Owner Worker Non-owner partner	Catamaram Units
2.	Individual Ownership Owner Worker Non-Owner Labourers	Catamaram Units
3.	Collective Ownership Owner Workers Non-Owner Labourers	Canoe Units Catamaram Units
4.	Collective Ownership Owner Non-workers Non-Owner Labourers	Shore Seine Units
5.	Individual Ownership Owner Non-Worker Non-Owner Labourers	Canoe Units Shore Seine Units

Source: Kurien,(1978)

As we go down the line the individual ownership is replaced by collective ownership. This is the dominant pattern of distribution of ownership in the traditional sector because when investment requirements are high, collective attempt to own the means of production among the fishermen is high. Canoe units and shore seine units require high investments compared with catamaram units. However, there are certain customary appropriation methods of the surplus in this sector which provides enough capital to make individual investments in costlier means of productions for a few. These successful fishermen often cease to be active fishermen and appropriate income as rentiers. This is evident from the fourth and fifth classification where owner non-worker is a class which is absent in other classifications. In the first three classifications, a distinct category of fishermen are owner workers. They are fishermen who can invest only in certain capital equipments by virtue of smaller investment requirement. This ensures an egalitarian distribution of ownership of fishing units of low value. It may be noted that the status of owner or owner-worker in the lower category of investment units is flexible. Fishery yields are uncertain. High returns lead to high accumulation and high investments in newer units or become partners in larger types of fishing units. On the other hand, when the returns are less, they may not be able to maintain ownership of the existing units and be forced to dispose of. Thus the status of owner and owner worker is very thin and exposed to vulnerability. This volatility in ownership is intense if the fishermen are indebted to money lenders.

Organisation of Work in the Traditional Sector

Organisation of work refers the division of labour or distribution of work among members of a fishing unit. The work organisation in the traditional sector varies according to the size of the craft. In a small catamaram unit, two or three persons share the work equally and interchangeably. In larger units the work is more complex and each undertakes a specified portion of the work.

In the case of canoe fishing, a work team of five members divide themselves into one steersman and four oarsmen. A general feature of the team is that all of them are proficient in fishing and seamanship. However, the steersman has an overall charge of the unit. Fishing operations are conducted under his directions and he is responsible to the maintenance of craft and gear. This responsibility gives him the power and prerogative to select the members of the crew. His skill requirement is of high order as to have a good knowledge of wind, weather, current, tides, habitat of fish and other relevant factors in fishing. He must know where to find the fish and must have 'good eyes' to locate the shoals of fish (Mathur, 1978).

Oarsmen are a major group in the fishing operations. They have specific duties to perform depending upon the place they occupy in the craft. Generally, in a plank-built canoe there are eight separate divisions extending from stern to prow (Mathur, 1978). These divisions which are separated with bamboo or wooden planks are known as Kallis (Compartments). These Kallis are called

1. Chukkan Kalli
2. Tamman Kalli
3. Idakalli
4. Kumbidi Kalli
5. Nalla Kalli
6. Mumba Kalli
7. Vittala Kalli
8. Komban Kalli.

The occupant of each Kalli has a definite and prescribed role during fishing operations.

In the fishing operations, normally the steersman shoot the net with the help of the man in the tamman kalli. This oarsman hauls up the net when the fish is entangled and puts the net back in to his kalli. The collegues in the idakalli and nalla kalli help him in shooting as well as hauling the net. In addition to this, the man in the nallakalli undertakes the duty to fix the sail according to the direction of the wind and remove it when the boat is rowed against it. The mumba kalliman sets the coir rope and anchor in position. The last two men near the prow take care of the catch and the nets and keep in safe custody the floats, sinkers, coir ropes, hammers etc. during the fishing expedition. They also assist the other oarsmen in shooting and hauling the coir rope, net, sinkers and floats. Thus in short, all the oarsmen co-operate with each other under the leadership of steersman in the fishing activity. The success of the fishing operation depends on the competence and team spirit of the crew (Korakandy, 1994).

It may be noted that in this sort of work organisation in the traditional sector, the work activity is within the control of real producers. The control of work activity along with lower level of investment enable the fishermen to have accessibility to a means of livelihood with out interruption, though productivity of such fishermen is low. Above all, this work organisation also ensures a fair share to each fisherman in the produce of the fishing efforts. In this regard, evaluation of the sharing system existing in the traditional sector is attempted.

A general rule in the matter of distribution of total share among the craft owners and the workers is the apportionment of the value at a given proportion taking into consideration both the interest of the craft owners and workers on the one hand and the reproduction of the system as a whole on the other. Depending upon the differences in the type of the fishing units, and the nature of ownership the proportion between the workers and owners may vary.

The Sharing System

In the case of fishing units with small investment, fishermen are likely to own more of productive equipments and they need not share the produce with any other. For example, in the case of small catamaram units, more fishermen families are able to own them and since their operation need not require any hired labour other than the supply of family labour, the output is fully appropriated by them. This is also the case with hook and line fishermen where the number of participants are limited and if they all belong to the same family. In such situations, there is no distinction between wages and profits. However, when investments in fishing units are larger, ownership will be limited and hence sharing of proceeds with hired labour is inevitable.

Table 4.2 shows the pattern of sharing existing under different craft-gear combinations in the southern part of coastal Kerala.

Table 4.2. The Sharing Pattern in Selected Craft-gear Combinations in Southern Kerala

Craft gear combinations	*No. of persons required to operate the craft*	*Rent on equipment (Percentage)*	*Workers share (Percentage)*
Small Catamaram hook and line	1-2	34	66
Double catamaram and net	4-6	25	75
Big catamaram and shark net	3-4	40	60
Big catamaram and drift net	3-4	40	60
Row boat and shore seine	20-40	40	60

Source: Vattamattom, (1978)

Information about the sharing system in the traditional fisheries of Kerala is also given by Kurien and Willmann (Kurien and Willmann, 1982). The scheme of distribution between the crew and owners of different craft-gear combinations are presented in table 4.3

Table 4.3. The Sharing Pattern of Different Craft-gear Combinations in the Traditional Sector of Kerala Fisheries

Craft gear combinations	*Owners share (Percentage)*	*Crew share (Percentage)*
1. Prawn nets with catamarams	50-60	40-50
2. Shore seines with dug-out canoes and sardine net with catamarams	40-50	50-60
3 Anchovy and large mesh drift nets with catamarams, hooks and line/ encircling net/shore-seine/small mesh drift net with plank built boats and large-mesh drift nets/lobster net with dug-out canoes	20-30	70-80
4 Cast nets with dug-out canoes and shore-seine with plank built boats.	10-20	80-90

Source; Kurien and Willmann, (1982)

The sharing systems show that in smaller units the share of the workers are higher than in the larger units. Similarly in units where the participants in the fishing process are larger, the share of the workers is larger. This shows that the 'livelihood' aspect of the work is evolved customarily in the traditional sector which is still remain as the basic norm of sharing system[1]. It may be noted that the share of the owners is more than forty per cent value if fishings unit is a costly one. Generally, it is from this share the capital accumulation emanates in the traditional sector. The owner-workers get an additional share besides the share they appropriate as owners of fishing crafts. This will augment their accumulation capacity. In the case of absentee fishermen, besides earning their income as rent, they render various other services to the fishermen. For example, the Karanavar or headman of a fishing unit organizes production in a fishing unit, by providing wage workers, purchasing equipments and arranging regular disposal of the catch. Such headman keeps account of the daily catch of the unit and settles the account of each-share holder. He also undertakes the responsibility to maintain the crew including the wage workers, during off-seasons. The headmen receive interest and commission for such services which augment their income (Korakandy, 1994)

A discussion about the accumulation aspect in the traditional sector requires us to look beyond the primary production. The primary producers and consumers are linked into an exchange relationship by a host of middlemen[2]. They include the auctioneers, the fish merchants, the money lenders etc. The people engaged in such functions may do their work distinctly or may overlap. Their economic surplus generally emanates from buying cheap and selling high. While discussing the pricing of the primary commodities, Robinson and Eatwell have hinted at the inevitability of middlemen and the economic advantage they get (Robinson and Eatwell). The primary producers are scattered, so also the final consumers and hence the inevitability of middlemen. Their financial leverage and oligopolistic position put both the primary producers and consumers under their maneuverability. This analysis holds true in the case of primary production of the fishery sector.

Interaction between the middlemen and fishermen result in subjugation of the traditional fishery by the former. The economic supremacy of the middlemen emerges from the peculiar nature of the primary production. Surplus production will not fetch much economic improvement for fishermen. If production is limited and market forces bring about a price increase, the advantage will only be reaped by middlemen. Thus at production level, the fishermen are at a disadvantage vis-a-vis the middlemen.

Further, the uncertainty of the primary producers with regard to their catch cause the same amount of uncertainty in their earnings. Coupled with the deprivation, during lean season, the fishermen are forced to depend on these middlemen to fend for poverty or to mend their damaged crafts. Middlemen exploit these situations by offering credits and secure accessibility to former's produce at terms favouring them. This readily fetters any improvement in the productive forces in the traditional sector making its growth very sluggish. This is clear when we look into the growth in the number of productive equipments and fishermen population during 1969-1982 (see table 4.4). For a systematic comparison we take the average growth rates for two periods (From 1969-71 and 1980-82)

Table 4.4. Number of Traditional Crafts and Fishermen Population

Year	*No. of crafts*	*Fishermen Population (Marine)*	
1969	29044		510553
1970	29560		523644
1971	30076		537070
Average of 1969-71	29560	Average of 1969-71	523756
1980	33448		617529
1981	33805		632967
1982	34162		648791
Average of 1980-82	33805		
Average of 1980-82	633096		

Source: Worked out from live stock census figures for 1972-77 and 1981, Govt. of Kerala.

A comparison of the average values between the two periods show that the traditional crafts have grown at the rate of 14 percent while the fishermen populations have grown by 20.80 percent. This shows that the growth of capital accumulation in the traditional sector lags behind the growth of fishermen.

Thus the labour process in the traditional marine fisheries is mainly motivated by a quest for making a livelihood for the participants[3]. However, surplus production was appropriated in the primary production mainly through non-economic means by certain individual fishermen who happened to possess more productive equipments and employed hired labour[4]. The inter-linkage of the fish economy by middlemen through the system of expanding markets brought some push to the productive activity in the sector but with in the control of these middlemen. Even though the middlemen have dominated the traditional marine fisheries, they were not able to disrupt the traditionally evolved equilibrium. The labour process was well with in the control of the real producers.

Labour Process in the Modern Sector

We have seen in the previous section that the development of productive forces in the traditional sector was hindered by the vested interests of middlemen. The domination by middlemen of both the product and credit market deprived the traditional sector of the 'surplus resources' that could have been used for its growth and expansion. Moreover, the formal intervention by the government for achieving 'progress' of the sector culminated in the intensification of this deprivation. The labour process evolved in the traditional sector for the development of marine fisheries on modern lines started in Kerala under the Indo-Norwegian Programme (INP) in 1953. This was in tune with the planned objectives of fisheries development formulated by Government of India. The salient features of the fisheries development under the first Five Year Plan comprised enlargement of the mechanised fishing fleet by motorisation of the existing traditional crafts wherever possible and by the introduction of new types of boats. The first plan has accorded the following priorities for marine fisheries development.

1. Mechanization of country crafts or introduction of new mechanised boats
2. Harbour facilities
3. Supply of requisites to fishermen
4. Development of marketing
5. Provision of ice and cold storage and transport facilities
6. Introduction of mothership operation and,
7. Provision of offshore fishing with large powered vessels such as purseiners and trawlers. Inland fisheries was also given importance

An examination of the planned objectives of fisheries development in the First Plan reveal that the objectives were premised on the presumption that the major constraint in the growth of fisheries was technological. It was earnestly believed that the rapid expansion of fishing crafts would lead to substantial increase in fish production and improvements in the living conditions of traditional fishermen.

At micro level, the objectives of the INP were :

1. to bring about an increase in the return from the fishing activity,
2. to introduce efficient distribution of fresh fish and improvement of the distribution of fishery products,
3. to improve the health and sanitary condition of fishing population, and
4. to raise the standard of living of the community in the project area, in general.

The INP attempted the modernisation process of the marine sector under an area approach. The Project selected two typical fishing villages in Kerala which covered an area of 25 sq. km in the extreme south-western part of India. These two villages were Sakthikulangara and Neendakara, situated 9 km north of Kollam town. The total population of the two villages was 11,157 of which one-third represented fishermen. The total number of fishing boats were 493, all were traditional crafts. Of these 493 crafts, 197 were big canoes and the remainder small canoes. The total investment incurred by these equipments were around Rs. 5 lakhs (Klausen, 1968).

The INP envisaged a phased programme of mechanisation to be implemented in three stages.

1. Mechanisation of existing crafts.
2. Mechanisation of small newly designed boats suitable for fishing with indigenous nets as well as new types of gear.
3. Mechanisation of bigger boats capable of staying out on the fishing grounds for longer periods of time as well as fishing in more distant grounds.

The project has skipped the first phase on account of the unsuitability of the traditional crafts to be mechanised. It developed a 22 feet clinker boat fitted with 4 to 5 H.P. diesel engine to begin with. No changes were effected in the gears. Traditional gill nets and drift nets were used. The successful performance of these boats induced INP to introduce other varieties of boats, superior in size and engine power. In 1959, the project introduced a 25 feet boat powered with 8 to 10 H.P. diesel engine. In 1961 a new design of 23 1/2 feet boat with 8-10 H.P. diesel engine was introduced.

In 1962, another category of boats, 25 feet 16 H.P. diesel engine designed specifically for operating shrimp trawl was introduced. Until 1963, the INP has constructed 150 mechanised boats. Between 1956-63, 138 mechanised boats were issued to the fishermen in the project area. The mechanised boats issued to the fishermen by size, are indicated in table 4.5.

Table 4.5. Type - wise Distribution of Mechanised Boats Issued to Fishermen, in Neendakara During 1956-63.

Year	*Types of Boats*				*Total*	*Cumulative*
	22ft (4-5H.P.)	*23 ½ft (8-10 H.P.)*	*25 ft (8-10 H.P.)*	*25 ft (16 H.P.)*		
1956	49	-	-	-	49	49
1957	18	-	-	-	18	67
1958	-	-	-	-	-	67
1959	-	-	9	-	9	76
1960	-	-	12	-	12	88
1961	-	13	15	-	28	116
1962	-	10	3	7	20	136
1963	-	-	-	2	2	138
	67	23	39	9	138	138

Source: Govt. of Kerala (1969), State Planning Board, Agriculture Division Studies – 4

An important aspect of this mechanisation process in the marine fisheries is that the INP attempted to equip the traditional fishermen with modern means of production. Through a system of subsidies and hire purchase facilities the price for mechanised boats were kept deliberately low. With as low an amount as Rs.1442/- one could acquire a mechanised boat during 1953-63. The subsidy given by Government to the fishermen were more than 50 per cent of the investment. The total investment incurred for the fishing equipments supplied to the fishermen were Rs. 1.03 million. The subsidised cost amounted to Rs. 0.55 million.

A disaggregated structure of the subsidy element given to the fishermen is shown in table 4.6.

Table 4.6: Distribution of Subsidies on Boats and Nets

Period	*Boat*	*Engine*	*Net*
Upto 1956	50.00%	50%	80%
1957-58	33.30%	50%	50%
1958-62	25.00%	50%	33%

Source : Klausen, (1968)

A number of studies were made about the economic impacts of INP on the fishermen communities. All these studies concluded that economic welfare of the fishermen community covered by the project has increased considerably. The result of INP project in terms of income appropriation by fishermen is shown table 4.7.

Table 4.7 : Economic Results of the Small Fishery Craft

(Measures are averages for 1959-62)

Sl. No.	*Particulars*	*Traditional Crafts*		*Mechanised Crafts*		
		Small canoe	*Big canoe*	*22ft (4.5 H.p.)*	*25feet (8-10 H.P.)*	*23'4ft (8-10 H.P.)*
1	Annual catch per boat (Kg)	8,603	8,010	8,841	14,563	11.652
2	Annual gross fishing income per boat (Rs.)	2,389	3,667	3,643	7,248	8,156
3	Annual net income per boat (Rs.)	2,032	3.148	1,883	4.559	5,244
4	No: of crew per boat	5	10	3 -4	4	4
5	Annual net income per crew (Rs.)	406	315	538	1,140	1,311

Source: Achari, (1962)

Table 4.7 reveals that annual fishing income from the mechanised boats is higher than the traditional boats. However, the difference of income between the traditional boats and mechanised boats narrowed down considerably when comparing the annual net incomes. This is due to the high operational costs of the mechanised sector. Again, the per capita income of the crew of the traditional sector and mechanised sector also differ much. This is because in the mechanised sector the net income per boat is shared by fewer number of people while in the traditional sector, the net income is shared by a larger number.

Under this modernisation attempt, the labour process has undergone tremendous changes. In this process, at the material level, new techniques have been incorporated. Definitely the new techniques are capable of producing more, taking advantage of the speed over time and space. Human labour is freed from the tiresome jobs. With less labour more came to be produced. However, these changes at the material level of the labour process missed a linearity. The changes have not organically emerged from the previous set of conditions but are super imposed (Platteau, 1985). To suit these changes at the material level of the labour process corresponding social relations are to be evolved. The new material conditions wrest from the labourer the accessibility to means of production because of the increase in the size of the organic composition of capital per unit of investment. Moreover, the change in the material conditions of the labour process intensified the characteristic of 'Open Accessible' nature of the fisheries. In short, the material changes in the labour process are such that they have altered the cost and character of investment. These 'cost' and 'character' aspects of investment, condition the social relations required to sustain the new labour process. It requires persons who are capable of and willing to make higher investments in fisheries. The emergence of such entrepreneurs depend upon the profit conditions in the sector.

However, the INP intervention delayed the formation of capitalist social relations typical of the capitalist labour process. The INP decided to supply the modern means of production to the real producers, the fishermen. This aspect prevented the possible separation of the labourers from the means of production at least temporarily. However, on account of heavy cost, the almost free supply of means of production could not proceed as an ongoing policy. It is natural that when such policies are withdrawn, the social formulation suited to the changed technological conditions will develop. The pace of such formations depend up on the profit opportunities in the sector.

One of the essential determinants of transition to full capitalist development is the availability of markets. Once, sufficient markets - internal and external- are opened capitalist production increases in quantity and circulation. During the initial period of planned development of the fisheries sector in Kerala, the market for fish products was nominal in extent. Of course, some individual attempts were made to export marine products, particularly frozen prawns in the 1950s[5]. However, it is the INP which proved to be supportive of the development of international markets for prawns[6] (Sebastian, 1986). The INP promoted exports directly and indirectly at three levels.

1. Resource confirmation and dissemination of the information regarding vast prawn beds in Kerala.
2. Introduction and development of bottom trawling using 30-36 feet trawlers.
3. Facilitating entrepreneurship in the form of inviting local fish merchants to make use of its freezing plants at a nominal rent for undertaking frozen prawn exports.

The production and export of frozen penaeid prawns during the period 1953-70 is shown in table 4.8.

Table 4.8: Production and Export of Penaeid Prawns, 1953-70 (Selected Years)

Year	*Total Production in Kerala (in tonnes)*	*Total Exports from India (in tonnes)*	*Share of total exports in total production*
1953	-	20	-
1956	14000	290	2
1957	20000	755	4
1960	13000	1843	14
1963	22000	6037	27
1966	28000	13367	48
1968	25000	21908	88
1970	37000	33684	91

Source: Sebastian (1986)

It may be noted that substantial changes have occurred in the exports of fishery products in the 1960s particularly after 1963. It is natural that these export boom attracted many entrepreneurs mostly non-fishermen. Entry of non fishermen in fisheries was facilitated because the technological composition constituted a new labour process in which capitalist had an upper hand.

Emergence of wage labour is a sine-quanon for successful capitalist transformation. In a sector, where traditional mode of production is dominant, availability of wage labour required for capitalist transformation can be ensured only by separating the common man off his means of production. This is a historical process of proselytisation of immediate producers into wage labour and the means of production and money into capital. In England, during Industrial Revolution, the formation of wage labour took place by driving away the peasants from their means of production viz. land. Thus, formation of wage labour by separation of workers from the means of production is an essential requisite for primitive accumulation [8]. Marx says;

"The so-called primitive accumulation, therefore, is nothing else than the historical process of divorcing the producer from the means of production" (Marx, 1978).

In Kerala fisheries, the conditions for primitive accumulation began to brew up with the inception of the INP. As noted earlier the INP pursued a policy of equipping the real fishermen with the new capital (mechanised crafts and gear). Because of the technical advantage and also because of lesser crew requirement to operate the new capital, labour productivity remained high, keeping the returns per fishermen also high. This is well documented by many impact studies of the INP programme. The economic impact of the INP in terms of income generation and appropriation is shown in table 4.7.

Table 4.7 makes it clear that the annual fishing income from the mechanised sector is higher than from the traditional boats. Even though the difference of income between the two sectors considerably narrows down when annual net incomes are compared, there is perceptible difference between the two sectors in the case of per capita income earnings of crew. Better income prospects in mechanised sector induce workers to migrate from traditional sector. Since both sectors are working competitively, expansion of the modern sector normally undermines the economic viability of the traditional sector. A secular expansion of the modern sector makes the owner workers in the traditional sector transform themselves into wage workers for the further enrichment of the capitalist.

Some of the structural transitions in the INP area are indicative of such changes. (See Table 4.9)

Table 4.9: Structural Changes in the INP Area (1953-63)

Nature of shifts	*Distribution of Households*		*Percentage change*
1. Entrepreneur households	1953	1963	
A. Non-operating (wholly with hired labour)	1.3	4.1	+215
(i) Mechanised boats	-	2.7	
(ii) Canoes	1.3	1.4	
B. Operating Households (with hired labour)	15.3	24.0	+57'
(i) Mecahnised boats	-	8.7	
(ii) Canoes	15.3	15.3	
C. Jointly Operating (Without hired labour)			
Q Mechanised boats	-	-	
(ii) Canoes	6.9	6.7	-

Source: Government of Kerala, State Planning Board (1969)

The economic and structural changes that have occurred in the INP area showed clearly the possible changes in the social formations of the marine sector. During 1953-63, the entrepreneur households who organised production wholly with hired labour and appropriated income from profits increased by 215 per cent. Similarly, another category of owners who operated alongside with hired labour also increased by 57 per cent. Thus it is clear that the successful performance of the mechanised sector in the INP area was on account of availability of wage labour. The supply of this wage labour was reinforced by separating the real fishermen from their means of production in the traditional sector. In the project area, it may be noted that, the fishermen who were operating their own canoes have declined by 3 per cent. Even though this was only a marginal change, it was a pointer to the permanent creation of a wage labour in the marine sector[9].

The opening of markets abroad, the making of a wage labour and favourable government policies have tremendously increased the accumulation potential of the marine sector. In other words, in the 1960s and 70s the marine fishery of Kerala poised for a full fledged capitalist development from the base of an accumulation brought in by the INP. We may further focus on the mellowing drives of capitalist relations in the marine sector of the Kerala economy by analysing the implications of labour process changes in the next section.

Implications of Labour Process Changes in the Marine Sector

The implications of the technical changes on the labour process in the marine fisheries of the Kerala economy can be summarised in terms of changes in the organisation of production, usurpation of work control by the capitalist through such organisational changes, the de-skilling of work activity of the fishermen, the development of intense capitalist relations reflected in higher production and productivity which in turn led to the deprivation and marginalisation of workers and finally the collapse of the marine sector due to its integration with metropole [10].

We have seen that in the traditional sector, the ownership of means of production was common with lower investment requirements. Conversely, where investment requirements were relatively high, such were satisfied by collective ownership. Under individual ownership mode only very few could become owners of costly crafts and gear. A professional group of entrepreneurs who were on the look out of profit opportunities seized of the profit potentiality in the marine sector. Non fishermen with financial prowess transformed themselves into captains of fishing industry (Platteau, 1985). The new technology served hand in glove with their commercial interest. The conditions of wage labour, as we mentioned earlier, and the availability of skilled persons trained by the INP strengthened their opportunities further.

The new super-imposed technology brought in its wings certain work changes. A new class of technically trained and experienced labour called 'shranks', ever willing to offer their services to entrepreneurs came up. Since fishing is by nature a co-operative work and reshaping of work activity by mechanisation enabled the shranks to emerge as a dominant players; availability of a team of surrogate workers to carry out co-operative fishing activity was to the advantage of shranks. This arrangement made a permanent mark up in the economic gain of the entrepreneurs because every attempt to increase efficiency by the fishing team assured the entrepreneur of 70 percent of additional income by the sharing arrangement existing in the sector. Besides, this arrangement saved the entrepreneurs of the trouble of finding workers to arrange the fishing trips.

The skilled activity of the shrank provide him some control over his workers. His superior status and the skilled nature of the work enabled him to appropriate a larger share allotted for the workers. In many instances, the share for the shrank is 40 per cent of the total crew share. In a five men crew, the rest, i.e. 60 per cent, is shared by the remaining workers equally. Thus it is obvious that under this arrangement, each unit of increase of work effort and consequent increase in productivity progressively benefit both the entrepreneur and the shrank vis-a-vis the workers.

The workers in the mechanised boats, barring the captain of the boat, are generally known as deck hands. They are expected to carry out the work orders given by the shrank. As such, the skill requirement by these deck hands are minimal as to be learned with in few days. We mentioned earlier that in the traditional sector, the work activity of each participant is skilled one and they learn it through experience from childhood. For the deck hands, there is no need of such learning-by-doing. In the early phase of mechanisation, the crew of the mechanised boats were drawn mainly from the fishermen community. However, at later stages due to de-skilled nature of fishing activity, crew members were from those outside the fishermen community [11].

"The capitalist mode of production systematically destroys all round skills where they exists, and brings into being skills and occupations that correspond to its needs" (Braverman, 1979).

The de-skilled nature of their work activity is reflected in assigning them only a small portion in the sharing pool. In a five member crew, after setting apart 40 per cent of the share to the shrank from the divisible pool allotted to the workers as a whole (30% of total value of the catch) remaining is shared by four members, thus each one getting just 4.5 percent of the total value of the catch. Even though these workers are entitled to only a small proportion of the total catch, in absolute terms, the sum they get would be high if the catch is high.

The de-skilled nature of work of the deck hands is reflected in the high labour mobility rate existing among them. In the mechanised sector, the degree of labour mobility is far higher than in the traditional sector. In fact, in the mechanised sector, an average boat crew changes the employer every 18 months (Platteau, 1985).

Thus we see in the organisation of production in the mechanised sector new social relations emerging from the technical alignments super-imposed up on the traditional sector. In this new social relations, the capitalist armed with the new technology usurp control over the labour process in the primary production of the marine sector and dominate the field.

It is technology that enabled the capitalists to split the work force into a group of skilled workers known as shranks on the one hand and another subservient group of deskilled workers- the deck hands. Through this fragmentation, control of the work is brought in to the hands of capitalists. By assigning a slightly higher share to the shranks, the capitalist wins the shranks to his side to get the work effort intensified. The shranks undertake the initiative to squeeze all possible work efforts to maximise the catch even though the major beneficiary of such work efforts is the capitalist entrepreneur. Thus the sub division of the work, which the technology permitted enabled the capitalist in controlling the work effort tightly. The creation of wage labour, and also the de- skilled nature of the work provided another opportunity for the capitalist to exercise their control over production. The labour market has turned into a buyers market because of the above reasons (Platteau, 1985). At the slight infringement of the production interest of the capitalist he has a free run in firing his workers, providing the capitalist an excellent opportunity of assuring continuous and uninterrupted flow of production with utmost efficiency. That the capitalist are not hesitant in invoking such powers is clear when we see that about 10 per cent of labour mobility is caused by such expulsions by the employers (Platteau, 1985).

Further, the accessibility of the capitalists to institutional and market credit[12] and their clout with political authorities in implementing policies in furtherance of their economic interests reinforce their control over work activity in the primary production of marine fisheries.

We may also look into the intensification of capitalist system of production of the mechanised sector via the growth in the share of its production. This sector had only a negligible share in the initial years of mechanisation during 1956-59. However, the commercialisation attempts of the INP paid dividends during 1960s and raised total production of the mechanised sector to 12000 tonnes. (See table 4.10) Since then, the favourable conditions for capitalist growth resulted in substantial increase in catch by this sector. The table 4.10 illustrates the trends in the marine fish production in the mechanised sector during the period 1956 to 1980.

Table 4.10: Trends in the Marine Fish Production in Kerala (1956 to 1980)

Period	*Total fish harvest (tonnes) (Average for period)*	*Share of mechanised sector*	*Percentage*
1956-59	2,37,000	Negligible	-
1960-66	2,88,000	12,000	4
1967-75	3,80,000	61,000	16
1976-80	3,32,000	1,02,000	31

Source: Kurien and Acbari ,(1988)

Table 4.10 shows that the percentage share of the total catch by the mechanised sector has increased several fold continuously over this period.

It is also interesting to look into the level of accumulation potential of the mechanised sector. The supply of wage labour which was almost perfectly elastic in the mechanised sector enabled the capitalist to retain the sharing system prevailed in the traditional sector. While in the traditional sector it is 70 per cent for the workers and 30 per cent for the owners on an average, in the mechanised sector this proportion is reversed. This sort of sharing practice in the mechanised sector enabled the benefits of mechanisation and productivity rise appropriated mainly by the capitalists. The trend in the distribution of total output value of fish in Kerala between workers and owners is brought out by table 4.11.

Table 4.11 : Distribution of Total Value of Output of Fish in Kerala Between Workers and Owners (Rs. Million)

Year	*Workers*	*Owners*	
		Of mechanised sector	*Of non-mechanised sector*
1969	103 (63)	19(12)	41 (25)
1974	293 (55)	143 (27)	99 (18)
1982	340 (45)	314(43)	88 (12)

(Figures in brackets show percentages)

Source: Kurien andAcbari, (1988)

The share from the total value of output by the mechanised sector has increased by more than two and half times during this period. Table 4.11 brings out the fact that the beneficiaries of capitalist development in the marine fishing sector were neither the owners in the traditional sector nor the workers in the traditional and modern sectors.

NOTES

1. The livelihood aspect is well ingrained in the traditional sector and it is still an important concern of traditional fishermen can be well explained with the Karanila system prevailing in coastal Kerala. Vijayan and Kurien have highlighted the concern of the fishermen about the community evolved mechanisms to ensure that the benefits from the common property fishing are spread as widely as possible in the whole community (Vijayan, Kurien, 1992).
2. The structure of marketing in the marine fish economy is discussed in detail By Kurien (1984).
3. Under Marxian theory this implies that the objective of production is making of use value.
4. The appropriation of surplus through non-economic means by these fishermen secured in different forms like service charges for some of the works rendered, interest charges for the money advanced, retaining a share as the owner of equipments etc.
5. Sebastian Mathew describes the attempts made by individuals to find markets abroad for shrimps particularly by one Madhavan Nair in the early 1950s (Sebastian, 1986).
6. Prawn fishery is the mainstay of the mechanised fishing industry of India (Kurien and Sebastian, 1982)
7. It is only from Kerala that prawns were exported substantially. Hence export of Kerala is almost equivalent to export from India.
8. Primitive accumulation of capital is the process by which the conditions necessary for the emergence of capitalism are created.
9. In 1960s and 70s no traditional canoes were found in the project area. They were all driven out of this area by the expansion of the modern capitalist sector and in the process converting the traditional owners into wage labourers.
10. A general contours of these type of changes are indicated by Baran. Discussing on the political economy of backwardness he remarked "But if western capitalism failed to improve materially the lot of the people inhabiting most backward areas,

it accomplished something that profoundly affected the social political conditions in underdeveloped countries. It introduced there, with amazing rapidity, all the economic and social tensions inherent in the capitalist order. It effectively disrupted whatever was left of the "feudal" coherence of the backward societies. It substituted market contracts for such paternalistic relationships as still survived from century to century. It re-oriented the partly or wholly self sufficient economies of agricultural countries toward the production of marketable commodities.It linked their economic fate with the vegaries of the world market and connected it with the fever curve of international price movements" (Baran, 1992).

11. It was pointed out that the composition of crew of some 22 mechanised boats which landed at Neendakara on a particular day found that out of 112 workers on board, only 26 or less that one quarter were traditional fishermen and among them 12 were deck hands. The skilled personnel like the shrank and the driver who get a large proportion of the crew share are drawn more frequently from non – fisherman community (Hakkim, 1980).
12. It has been found that since the credit worthiness of the entrepreneurs in the mechanized sector are high, they were able to secure loans at lower rates of interest from the professional money lenders (Platteau, 1985).

REFERENCES

1. Achari, T.R.T., (1962): *Economic Assessment of Fishing by Small Mechanised Boats and Canoes in the INP Area,* INP, Quilon.
2. Achari, T.R.T., (1962): "Indo-Norwegian project: An Assessment of Its impacts on the Growth and Development of Indian Fisheries", *Souvenir,* INP, Quilon.
3. Bagchi, Amiyakumar, (1984): *Political Economy of Under development,* Oxford University Press, Delhi.
4. Baran, Paul, (1992): On the Political Economy of Backwardness", *in Political Economy of Development and Under development,* Wilber. K. Charles and Jameson, P. Kenneth, MC Graw-Hill.
5. Department of Animal Husbandary, (1973): Livestock Census - 1972, Govt. of Kerala, Trivandrum.

6. Hakkim. A.V.M., (1980): *Mechanisation and Co-operative Organisation, Their impact on Traditional Fishermen of Kerala* (Unpublished M.Phil Dissertation), Centre for Development Studies, Trivandrum.
7. Klausen, A.M., (1968): *Kerala's Fishermen and the Indo-Norwegian Project.* Universitet Forlayet, Oslo.
8. Korakandy, Ramakrishnan,(1994) : *Technological Change and the Development, of Marine Fishing Industry in India.* Daya publishing House, New Delhi.
9. Kurien, John, (1978): *Towards an Understanding of Fish Economy of Kerala State,* (Working Paper Number 68), Centre for Development Studies, Trivandrum.
10. Kurien, John, (1984): *The Marketing of Marine Fishermen, Inside the Kerala State - A Preliminary Study* (Mimeographed Report) Centre for Development Studies, Trivandrum.
11. Kurien, John, and Sebastian, Mathew, (1982): *Technological Change in Fishing Its Impact On Fishermen,* Centre for Development Studies, Trivandrum.
12. Kurien, John and Willmann, Rolf (1982): *Economics of Artisanal and Mechanised Fisheries in Kerala - A Study of Costs And Earnings of Fishing Units,* (Working Paper No.34) FAO/UNDP, Madras
13. Kurien, John and Achari, T.R.T., (1988): "Fisheries Development Policies and Fishermen's struggle in Kerala". *Social Action,* Vol. 38.
14. Govt. of Kerala, (1979): *Livestock Census 1977,* Trivandrum.
15. Govt.of Kerala (1984): *Livestock Census-1982*, Trivandrum.
16. Marx, Karl, (1978): *Capital Vol. I*, Progress Publishers, Moscow.
17. Mathur, P.R.G., (1978): *The Mappila Fisherfolk of Kerala,* Kerala Historical Society, Trivandrum.
18. Platteau, J.P. et. al., (1985): *Technology and Credit Indebtedness in Marine Fishing . A Case Study of Three Fishing Villages in South Kerala.* Hindustan Publishing Corporation, India.
19. Robinson, J: *An Introduction to Modern Economics,* Tata Mc Graw Hill, New Delhi.

20. Sebastian, Mathew, (1986): *Growth and Changing Structure of the Prawn Export Industry in Kerala, 1953-83* (Unpublished M.Phil Dissertation), Centre for Development Studies, Trivandrum.

21. State Planning Board (1969): *The Impact of Indo-Norwegian Project on the Growth and Development of Indian Fisheries,* Agricultural Division, Studies, State Planning Board, Kerala.

22. Vattamattom, Joseph, (1978): *Factors that Determine the Income of Fishermen. A Case Study of Poonthura Village in Trivandrum.* (Un published M.Phil dissertation) Centre for Development Studies, Trivandrum.

23. Vijayan, A.J., and Kurien, J., (1993): "Symposium on Socio-Economic Issues in Coastal Management", *IPFC. Bangkoke. Thailand, November.*

5

Labour Process Changes : Conflicting Terrains and Labour Class Responses

The labour process changes in the fishery sector of Kerala covered by chapter four was the story of the success of the capitalist in obliterating the organically evolved labour process in the traditional sector with alien technology. Capitalist intrusion served to toss coastal Kerala into the vortex of conflicts, yet another epoch of the theory of labour process - the contested terrain. This chapter focuses on such conflicting terrain.

Marginalisation of Fishermen Community and Fish Workers

The saga of modernisation of marine sector in Kerala is also one of marginalisation and peripheralisation of the artisanal fishing community. Until the modernisation attempt, the capture fisheries was exclusively the preserve of the fishing communities. There was no external threat to their economic domain, but their was nothing to write home about their economic status. Governmental efforts to improve their lot stressed the need to raise their productivity. The early attempts of the Government were on a realistic approach of leaving fisheries as a source of occupation and livelihood of a considerable section of the population of the state for a long time (Kurien and Achari, 1988).

Based on these facts, a two fold approach was resorted to: first to upgrade the existing traditional technologies and then gradually introduce new ones. By the former strategy efforts were made to supply wood for traditional boats, cotton for nets and the setting up of curing yards for hygienic processing. Introduction of nylon nets, issuing the fishermen with small mechanised gill-net boats and establishment of more ice plants were the chief components of the latter strategy. Some institutional arrangements like co-operatives were also made to disseminate the new technology to reach the real producers. The skill and accumulated knowledge of the fishermen were considered crucial in imparting complex skills unique to the modern technology.

However, a continuum of this realistic approach was considered as irrational and consequently neglected when viewed, against the 'new paradigms of developments based generally on the imitation of western models. Rapid modernisation with big strides in technology vertically imposing upon the existing structure was found to be the key approach under the new development paradigm. The export possibilities added verve to this approach. Thus we see in Kerala fisheries two phases of modernisation: first a slow phase and then a rapid phase (Kurien and Achari, 1988). The transition from a slower phase to a rapid one has resulted in economic progress in fisheries of Kerala, but has rendered the artisanal fishermen sapless in the following ways.

Table 5.1 : Trends in Marine Fish Production in Kerala State (1956-1980)

Phases	*Periodisation*	*Total fish harvest (000 tonnes) (Average for period)*	*Harvest by technology(000 tonnes)*	
			Non-Mechanised	*Mechanised*
Slow modernisation Phase	1956-59	237	237(100)	
	1960-66 (Use of nylon nets and gillnet boats)	288	276(96)	12(4)
Rapid modernisation Phase	1967-75 (Use of trawlers and trawl nets)	380	319(84)	61(16)
	1976-1980 (Use of more trawlers and purse seiners)	332	230(69)	102(31)

Source: Kurien and Achari, (1988) (Figures in brackets show percentage)

1. By rendering their means of production less productive and non-viable in the context of the 'new progress' that was unfolded in the fishing sector[1].
2. By forestalling the artisanal group from receiving due share in the organized development attempts of the Government.

The figures that follow shed light upon the progressive marginalisation of the artisanal fishermen during the modernisation phase. We may first look into the trends in fish production.

It is clear from table 5.1 that over the period of twenty years (1956-1975) the catch has increased. But the fruits of production enhancement were not shared uniformly. In other words the traditional fishermen were not the beneficiaries from such a growth. During the slow modernisation phase, because of the realism in the government policies, in the first decade from 1956-1966, the benefits of growth were confined among the real producers. During the 1956-59 period, 100 per cent of the total catch was contributed by the artisanal fishermen. When nylon nets and small gill net boats were experimented by the artisanal fishermen, the production has increased and the artisanal fishermen were the sole beneficiaries of such growth. However, that only a small number of fishermen were successful in becoming owners of the mechanised means of production with the aid of government schemes and their contribution to the total catch was just four per cent during this period. Even though, slow in pace, the new set of conditions with appropriate technological changes contributed to a measurable increase in productivity of the active fishermen. The fish catch per fisherman per annum rose to 3,800 kg in 1965 from a 3500 kg in 1961. The per capita income of the fishermen on an annual basis also rose from Rs.330 in 1961 to Rs,540 in 1965 (Kurien and Achari, 1988).

However, the new shifts in policy approach did not give ample time to mature and dissipate the above set of conditions in the entire artisanal fishery. On the other hand, it boosted the commercial interests resulting in penetration of capitalist interest particularly among the outside capitalists who were remotely connected with capture fisheries. The rapid modernisation phase with the use of bottom trawling technology enhanced production tremendously during the period 1967-75. The average production during this period has increased to 380000 tonnes, an increase of about 32 per cent when compared with the previous slow modernisation phase. Of this increase, more than 53 per cent of the growth was the share of the mechanised sector, whereas the non-mechanised sector contributed only 47 per cent. In terms of the growth per annum, during the phase of rapid modernisation the growth rate was more than 45 per cent in the mechanised sector and about five per cent in the non-mechanised sector. The mechanised sector growth was spectacular but in the non-mechanised sector the growth rate of the previous slow modernisation phase dropped from 2.7 to 1.7 per cent per annum.

The second phase of the rapid modernisation too intensified the marginalisation and peripheralisation of the artisanal fishermen. Despite the introduction of new trawlers and purse-seiners, the fish production as a whole has declined. The overall fish landings, on an average, has declined to 332,000 tonnes from 380,000 from the previous period, (a decline of about 13 per cent). It may be noted that this decrease in the level of output has not affected the mechanised sector. In fact, their fish landings have increased from 61,000 tonnes to 102,000 tonnes (an increase of more than 67 per cent). The decline in the total fish landings was on account of the fall in catches of the artisanal group. In fact, their production fell to 230,000 tonnes, from the previous high of 319,000 tones. The waxing of the mechanised sector and the waning of the traditional sector is well exposed in the last columns. In terms of annual growth rate, while the mechanised sector has grown by more than 13 per cent, the artisanal sector registered a decline in production by 5.6 per cent.

In the backdrop of an annual growth rate of fishermen population of roughly two per cent, modernisation of the fishing sector has worsened the economic conditions of the artisanal fishermen, with a growth rate of production of 1.7 per cent in the first phase of modernisation and a negative growth of 5.6 per cent in the second phase.

The artisanal fishermen advance specific reason for decline in their catch. Generally, the mechanisation drive has resulted in ecological damages the onus of it is greater upon the traditional fishermen. Indiscriminate trawling and consequent obstruction of fish breeding systems have upset the eco-system causing drastic fall in fish population which accounts for drastic reduction of the average catch, size and earnings of fishermen.

The traditional fishermen argue that their catch has been declined due to three specific reasons.

1. The raking of the sea by the bottom trawlers results in damage of fish eggs and larvae and disturb nursery ground of fish. The artifacts like trawl nets scoop up a lot of juvenile fish.

2. The noise of the mechanised boats frighten off fish and the operation of the boats particularly during night times results in damaging the nets of the artisanal fishermen by the propellers of the mechanised boats.
3. Trawling operations cause turbidity of the sea and fish shoals avoid the muddy sea and escape their gears (Kurien and Mathew, 1982).

The competition at two levels in the production sphere also adversely affected the traditional fishermen. First is the competition for space. Trawlers are in pursuit of demersel species while non-mechanised crafts with their drift nets primarily intend to catch larger migratory pelagic fishes. While they fish in the same area, they are infact, competing for space and generally the latter always incur huge loss of capital.

Similarly, while the mechanised sector resorts to chasing the fish shoals and encircling them enmasse, the traditional fishing wait for shoals to enter into their ambit of operation. These two techniques known as active and passive fishing contest each other in the same area, the latter would be at a disadvantage resulting in lower catch and income (Kurien and Mathew, 1982).

The experiences of ecological damages and the consequent decline in catch is well known to the fishermen 'with their intimate familiarity with the sea'. It is also corroborated by scientific community even though the issues involved have not been conclusively settled.

The fishermen sauvy of the sea and fish acquired through their constant familiarities with the sea makes the damages to the eco-system and consequent decline in fish population so close to their heart. A U.N. study clearly establishes a relation between noise generated by the fishing vessels and fright reactions of fishes. The study shows that fishes had a violent escape behaviour by diving and avoidance of the disturbed area by oriented changes in swimming direction (FAO, 1979).

Similarly, it has been well established in fisheries biology that inshore waters form the nursery grounds of all kinds of fish. Gulland has pointed out "it is an unfortunate biological fact that the nursery grounds tend, when they exist, to be comparatively close inshore" (Gulland, 1979) The reason given is that the lower salinity level in the inshore region provide some security to the little fishes as their predators cannot withstand the lower salinity condition. Given this biological fact, there is evidence to the fact raised by the artisanal fishermen that trawlers adversely affect the breeding grounds.

The marginalisation of the artisanal fishermen were not confined to attempts at eradicating them from the sea and depriving them of fish but also by diverting the public funds away from projects which would benefit the children of the sea. An examination of the plan expenditure incurred for the development of marine fisheries during the slow modernisation and rapid modernisation phase reveals that the artisanal fishermen were not the principal focus in the development schemes initiated in the fisheries sector during this period [2]. The plan expenditure incurred in the Kerala Fisheries during 1956-1980 is shown in table 5.2.

Table 5.2 : Kerala Marine Fisheries - Plan Expenditure (1956-1980) ***Rs. in million***

Sl. No.	*Item*	*Slow modernisation phase (1956-1966)*	*Rapid modernisation phase (1967-1980)*
1.	Upgrading traditional craft	2 (5)	6(2)
2.	Introducing new craft	11(27)	102 (36)
3.	Fishery infrastructure	8(19)	27(9)
4.	Processing and marketing infrastructure :		
	- Oriented to internal market	3(6)	2(1)
	- Oriented to export markets	8(21)	9(3)
	- For fisheries development corporation	-	46 (16)
5.	Credit for fisheries co-operatives	2(5)	19(7)
6.	Training schemes for fishermen	1(3)	6(2)
7.	Welfare measures, social infrastructure	1(3)	60 (20)
8.	Administration	2(5)	5(2)
9.	Other items	3(6)	6(2
	Total expenditure	41	288

Source: Kurien and Achari, (1988) Figures in brackets are percentages.

Even though during the rapid modernisation phase the total expenditure has increased from Rs.41 million to Rs.288 million the allocation made towards the direct improvement of the artisanal fishermen has not shown any reasonable increase. Infact, allocations that would have improved the conditions of the artisanal fishermen has declined during the second phase of modernisation. For example, five percent of the allocation had been given for upgrading traditional craft during the slow modernisation phase. It declined to two per cent in the second phase affecting directly the artisanal fishermen. Similarly, a one per cent cut from the training schemes of fishermen during the slow modernisation phase affected the traditional fishermen adversely. Of course, there is an increase of two per cent in the credit given to fisheries cooperatives but it is already an established fact that the benefits of increasing credit have been usurped by the well off fish merchants rather than the artisanal fishermen (Hakkim, 1980)[3]. A cut in the allocation earmarked for processing and marketing infrastructure particularly in the development of internal markets directly affects the artisanal group. An increase of 17 per cent is seen to have occurred during the rapid modernisation phase in the case of allocation to welfare measures and social infrastructure. But there is no proof to show that such an increase has led to an enhancement of the productive ability of the artisanal group. The only reality is that they were cornered by the 'new modernisation schemes'.

The allocation of funds by the private sector also favored the capitalist class owing to their credit worthiness and solvency. The artisanal fishermen are shown the door by private and public organised credit agencies. Naturally they are thrown off to the mercy of the money lenders giving the mushrooming exploitative practices (Platteau, et al., 1985).

The pathetic existence of the traditional fishermen was put out by the socio-economic census conducted by Department of Fisheries in 1979. The fishermen were low in income and educational attainments, housing conditions were poor with 48 per cent having only dilapidated huts, access to drinking water was limited, sanitary and lighting facilities were abysmally low. This squalor, poverty and deprivation and uncertainty about future provided a breeding ground for dissent and protest. The restlessness of the fishermen caused by their marginalisation and peripheralisation engendered by the avarice of the capitalist began to crystallize into collective actions. Similar to the fundamental opposition of the working class against the capitalists, sharp responses emerged from the real fishermen but in novel ways due to the specific relations in which the marginalised group have put themselves in and also against the capitalist class in LDCs. We may now delve into the "constructive responses' initiated by the fishermen against their plunder and exploitation by the capitalist intruders.

The responses evolved from the fishermen community were mainly in two directions.

1. Strengthening their bargaining power as a class against all the forces which were instrumental in marginalising and peripheralising them, through collective actions.
2. Collective and individual attempts to improve their means of production by incorporating viable alternative techniques. We may first focus on the former.

The fishermen of Kerala were not cohesive as the name suggested. There were no general platform to bring them together for a long time. Moreover, the fishermen belonged to Hindu, Christian and Muslim religious groups and were subject to their religious dogmas and restrictions and thus remaining isolated groups. They were unable to cut across such religious divisions to understand and identify the common economic interests. However, there were some organisational attempts to unite the respective groups on religious lines without any explicit economic motives behind. Two such associations were Dheevara Sabha and Latin Catholic Associations. Their main concerns were social and familial issues of the respective communities (Ibrahim, 1986).

However since independence political parties made certain attempts to organize fishermen on account of vote bank considerations. Both the congress and communist parties vying each other in this pursuit. The congress party focused on the anti-communist feelings of the fishermen enclaves who were traditionally grouped on religious lines (Ibrahim, 1986). The communist party, on the other hand intruded into all possible areas of fishermen enclaves impressing them of the need for collective action in safeguarding the rights of the fishermen in the wake of modernisation process. Attempts at organising fishermen during 50s and 60s taught that they would not come around on a common platform of fishermen needs (Ibrahim, 1986); for the fishermen remained complacent.

Initially, the traditional fishermen were the focus of modernisation and the thrust was on progressively equipping them with the modern means of production. However, all such objectives which aimed to keep the traditional fishermen in the central place of modernisation failed. The modernisation which assumed capitalist line of growth progressively marginalised the traditional fishermen and deprived them their only means of production. This fact induced the fishermen of all the caste and religion to come together to recover and preserve the fishing ground lest to be wiped out.

However, the pattern of collectivization and unionization of fishermen were region and situation specific. In the mid-seventies Christian clergies who had abiding interest in fishermen cause started conscientizing them of their plight in the wake of mechanisation. This created in them a new awareness, a sense of seriousness and immediacy, which induced them to organise themselves in trade unions to find ways and means to avert a disaster. This resulted in the formation of small fishermen unions in various coastal districts of the state.Table 5.3 shows the progressive efforts of fisher folk to organise themselves into unions.

Table 5.3 : Formation of Fishermen Unions in Coastal Kerala During 1970s and 1980s

Sl. No.	*Name of the union*	*Formative year*	*Place*
1.	Alappuzha district fish workers union	1970	Alappuzha
2.	Marine fish workers union	1970	Kochi
3.	Alappuzha catholic fish workers union	1970	Alappuzha
4.	Vijayapuram parish fish workers union	1977	Vijayapuram
5.	Thiruvananthapuram parish fish workers union	1978	Thiruvananthapuram
6.	Anchuthengu boat workers union	1978	Anchuthengu
7.	Thiruvananthapuram district fish workers union	1979	Thiruvanathapuram
8.	Cochin area fish workers union	1979	Kochi
9.	Malabar independent fish workers union	1982	—
10.	Ernakulam district fishworkers union	1982	Ernakulam

Source: Kaleeckal, J. (1988)

The ambit and the issues of these unions remained local. When the unions were initially formed there could not be any articulative direction in regard to resolving the fishery crisis since they could not grasp the dialectics of the capitalist growth that got entrenched in the fishery with modernisation attempts. Hence they got themselves involved in seeking immediate and short run relief like charities and favours from the government and other agencies in the form of free ration, demand for sea wall construction, settlement of displaced fishermen etc.

The growing awareness of the root cause of their problem and the experience of success[4] of their collective efforts at various levels helped them to form a state level organization known as Latin Catholic Fish Workers Union in 1978. Shedding its religious and geographical overtones, it grew up as Kerala Swathanthra Matsya Thozhilali Union (Kerala Independent Fish Workers Union) in 1980.

"The Kerala Swathantra Malsya Thozhilali Federation is a federation of different district level trade unions of small scale, fishermen belonging to all caste and creed. It is the most secular form of organisation, where Muslims, Hindus, Christians, Priests and Nuns collaborate and struggle together for the betterment of the traditional fisher folk" (Kochery, 1982).

All the living struggles of the fisher folk were conducted under the banner of this union since then[5].

With the formation of a state wide union the fishermen struggle underwent a qualitative change. Strategically, they adopted more and more of intervening and proactive tactics. The union focused its activities on two lines. At one level, it tried to give a scientific temper for the traditional knowledge base by organising awareness programmes. The services of voluntary organisations and individuals who were sympathetic to the cause of traditional fishermen were also made use of. At another level, it focused on increasing the socio-economic awareness particularly in the context of the dialectics of capitalism.

The height of their increasing consciousness and the optimism in their collective efforts prompted them to float 'The National Forum for Catamaram and Country Boat Fishermen's Rights and Marine Wealth'[6] for more rigorous struggles.

Thus we see that the traditional fishermen who were a fragmented lot for a long time got organised in the decade of 70s and early 80s as a strong force determined to challenge the capitalist intrusion in their domain — their only source of livelihood. So far, we have described the political response of the traditional fishermen and now we may focus on the technological response which is considered to be a peculiarity of traditional workers in the socio-economic context of LDCs.

Technological Responses Catering Artefactual Improvements

The artisanal fishermen developed their artefacts overtime in an informal set up. However, in the face of capitalist development such 'gradual approach' of technical improvements became insignificant (Kurien, 1990). Under capitalist production system commercial interest is the prime determinant of technical changes vis-a-vis the subsistence and sustainable motives of technical improvements under traditional sector. Thus the dynamics of technical developments occur at two different levels affecting people differently. In this sense, technical changes have certain class dimensions. The agencies associated in the development of technology and the motives attached in such developments further prove this dimension. Certain altruistic Non-Government Organizations and individuals had the sensed the possible marginalization of the traditional fishermen because of capitalist intrusion. They felt the necessity of resisting such marginalization by increasing the productive ability of the artisanal group. Development and incorporation of a new technical packages involving plywood boats (PBs) and out board motors (OBMs) were some certain crucial steps in this regard. The principal agencies associated with the development of a viable technology in the traditional sector were the Kottar Social Service Society (KSSS), a social organisation working among fishermen in Kanyakumari District of Tamil Nadu, Fishermen Community Development Programme (FCDP), working among the artisanal fishermen in Quilon District in Kerala. South Indian Federation of Fishermen Societies (SDFFS) an apex federation of fishermen's organisations and a host of individual fishermen. They were also helped by the Intermediate Technology Group - (ITDG) the appropriate technology centre set up by the famous economist Schumacher (Kurien, 1994).

Making New Crafts

In 1972, Fr.TJames, Director of KSSS began a search for boats affordable to fishermen. His idea of making a fibre reinforced plastic boats (FRP boats) resulted in inviting a Belgian electro-mechanical engineer, Pierre Gillet. After his arrival in 1973, KSSS started a training programme in FRP (Fibre Reinforced Plastic) moulding in its new Boat Building Training Centre at Muttom. However, some of the models developed were uneconomic when considered against the economic condition of the fishermen. Focus was given for construction of boats affordable to fishermen. Gillet formed a Centre for Appropriate Technology (CAT) and geared to finding technological solution to a nascent need of the artisanal fishermen of the region. In this endeavour the CAT got the assistance of Gifford of Intermediate Technology Development Group (ITDG). By the end of 1981, a prototype was made and successful trials were conducted and the model came to be known as 'Muttom Cat'. Renewed versions of the model suited to the ecological and economic conditions of fishermen were developed and this was popularised as the 'Kottarkat'. The most striking feature was that the cost of the boats were well within the reach of the fishermen [7].

In 1982, the FCDP at Kollam District wanted to replace their traditional dugout canoes. Their initiative and contact with CAT helped them in the development of 'Lakshmi Vallams' or plywood valloms. Thus we see that certain NGOs and individuals along with the efforts of fishermen provided them a technical edge to compete the mechanised capitalist sector by developing efficient and cost effective crafts.

Fixing Outboard Motors

Adoption of outboard motors turned out to be another technical leap forward. It may be noted that motorisation of country crafts was ruled out as technically impossible by INP and adopted foreign models' suited to coastal conditions in the state. This aspect has not distracted the artisanal fishermen and their allies in their search for finding alternatives against mechanisation. In 1969, the KSS tried to motorise 100 catamarams by importing powerful 18 H.P. Evinrude petrol /kerosene engines imported from Belgium[8], However, the project was wound up in 1973 and it has been pointed out that "The failure of the experiment cannot be attributed to any single factor but to a chain of adverse circumstances". Further, it was pointed out that "technical dependence on foreign skill and equipment was certainly a major handicap" (Gillet, 1979)[9]. In 1974, the Marianad Malsya Ulpadaka Co-operative Society in Thiruvananthapuram District tried to introduce OBMs but failed[10]. It was followed by another experiment in Purakkad, with the help of Kerala Fishermen's Welfare Corporation established in late 1980 by Government of Kerala. Another group of fishermen in Ernakulam District successfully conducted transformation of their crafts to fix OBMs (Alagarajan, 1994). In short, the increasing marginalisation and peripheralisation forced many individual fishermen to remodel their crafts and adapt OBMs to venture into sea as a life and death matter.

Fish Attracting Lanterns (FAL)

Another, survival strategy introduced by the marginalised fishermen in the form of a technical improvement was the use of FAL. Of course, this practice is not a new method in fishing but in marine fisheries in Kerala it is a novel feature. The use of lanterns as a method to attract fish has been in existence in many countries like Japan, China, France, Russia, Philippines, Korea and Thailand etc (Rajan, 1995). In Kerala lanterns began to be used as part of a technical package in the fishing process in 1985. The FALs are used during night time fishing by fishermen using hook-and-lines and boat seine methods. Improvements within this technique have occurred since then. To begin with kerosene lanterns were used but were replaced by petromax lights in 1987. The use of gas could not succeed in the operation[11]. The popularity of this method could be guaged from the pace of its spread among fishermen. The fishermen of Marianadu in Thiruvananthapuram district used it for the first time and it spread to 27 out ofthe 47 coastal villages of the district (Rajan, 1995)[12]. The number of lanterns used in each fishing units was four or five at the beginning, which has increased to six or seven, later.

Peoples Artificial Reefs (PARs)

Marine resources are 'common property' resources and hence are 'open access' resources. Generally they are open to use by all and owned by none. However, the accessibility to fish harvesting was restricted by the skills required to appropriate from the common base. The community of fishermen have over generations acquired and transmitted such skills through learning-by-doing, and transmitted such knowledge to subsequent generations. Alien capitalist technology and techniques enabled non-fishermen capitalists to rip through the community skills and plunder the resource base for profit. The wanton destruction of the resource base was inconsequential to the capitalists. The 'common property' of marine resources being the only source of livelihood for the fishermen community over centuries, the destruction of the same could not be viewed with apathy. Rather, determined, though sporadic, measures to restore and rejuvenate this common property resources camp up. Along with these micro measures certain macro measures were also evolved to regulate and safeguard the resource base. First we may deal with the micro measures adopted by the fishermen to restore and rejuvenate their resource base. The fishermen at village level assembled to rejuvenate their fishing field through construction of artificial reefs. New reefs were developed primarily as a result of the initiative of the hook and line fishermen who were convinced that they had to help Kadalamma (Mother Sea) to rejuvenate herself after the onslaught of trawlers.

"Coastal fishermen live at a particular spot on the coast for generations and thus have thoroughly mastered the topography of the inshore waters, the profile of the sea current and other hydrological fluctuations and the related fisheries so that they constitute an 'eco-society' that has ecologically turned itself to the coastal eco-system that they have been living in. They have learned to live in perfect harmony with their coastal environment to conserve their natural resources and even to manage them so judiciously as to be reckoned as the self appointed custodians of their coastal eco-system" (Raj, 1990).

It is this life-oriented view that induced the fishermen community in the context of capitalistic led marginalisation to unite at various micro levels to form a socio-ecological movement at macro scale.

It is the experience of the traditional fishermen that external objects in the sea attract fish [13]. This prompted them to place artificial reefs to attract fish, provide or improve fish or shellfish habitat and increase fish biomass locally. Attempts to augment fish by increasing artificial reefs remained very limited owing to the relative abundance of fish in the early phase of modernisation. However, it was the marginalisation of the traditional fishermen that led to the resurgence of interest in artificial reefs. The reefs so fabricated, came to be known as 'People's Artificial Reefs' (PARs)[14]. As many as 22 villages, took part in the collective action of erecting PARs in the coastal waters of their villages (Kurien, 1990). During the first decade of the fishermen movement (1979-89) the pace at which PARS were erected increased substantially. The table 5.4 shows the artificial reefs erected during the period between 1960-89

Table 5.4. Pace of Construction of PARs in the Post-Motorisation Phase

Period	*Before 1960*	*1979-83*	*88-89*
No. of PARs erected	2	9	21

Source: Kurien, (1990)

The table 5.4 shows clearly that since 1975, the period of marginalisation,the number of PARs have increased rapidly. Thus we see that the traditional fishermen responded against the capitalist intrusion and the subsequent marginalisation and deprivation not only in the form of organising unions and asserting their rights but more constructively by improving their means of production incorporating appropriate technology through developing it and in certain cases adapting it. This is a novel reaction of the marginalised community to fight against the capitalist intrusion and derangement of their domain of livelihood.

The above narrated responses of the traditional fishermen by adopting technological changes and organising themselves into unions, ushered in a specific phase since 1980 and came to be known as 'motorisation phase'. We consider the motorisation phase in fisheries development in Kerala as an attempt of the traditional fishermen to recover control of fisheries sector from where they were squeezed out for more than two decades in the 60s and 70s.

NOTES

1. The specific ways through which the means of production and efforts of the traditional fishermen were rendered nonviable in context of capitalist development is explained in chapter four.
2. A review of plan progress in 1993 reveals that out of the total plan expenditure on fisheries of Rs.39 crores, since the introduction of mechanisation upto the end of Fifth Five Year Plan, 79 per cent had gone to the development of mechanised fishing (Government of Kerala, 1993).
3. The institutional arrangement to equip the fishermen with the modern means of production was co-operativisation but it failed because such co-operatives were hijacked by vested capitalist interests (See Hakkim, 1980).
4. In Trivandrum district, at Anjengo, the formation of Boat Workers Union in 1977, enabled them to force the government to stop the revenue recovery proceedings against the fishermen who happened to be the recipients of supply of mechanised boats which were technically defective and operationally inefficient causing huge losses to the fishermen. In Kollam and Alappuzha districts also they succeeded in conceding many of their demands by the authorities.
5. The details of various strikes strategies and programmes organised by the fisherfolk through their collective efforts is described in 'Oru Samarakatha' (A Story of Struggle) by Jose. J, Kaleeckal and others (Jose Kaleeckal, et.al, 1988).
6. This fishermen's Forum was formed in July 1978 at Delhi where 13 associations representing traditional fishermen from different parts of the country assembled.

7. At 1982 prices, the cost of a boat was Rs.7500 only.
8. The attempts of KSSS to motorise the traditional crafts in 1968 was started with the help of Indo-Belgian Fisheries Proejct (IBFP) and Freedom from Hunger Campaign (FFHC) Delhi and other donor agencies. Their activities were mainly confined in three areas.
   (i) Introduction of nylon nets,
   (ii) Mechanisation of catamarams, and
   (iii) Tests on beach - landing crafts (Gillet, 1979).
9. Even though this attempt has failed in securing its objectives, it has indicated the scope of an organic and a linear growth in the fishery (See for details of the project KSSS, 1971). However it has not materialised because of the over enthusiasm with modernisation attempts.
10. For details of this experiment of fixing OBMs in canoes see Kurien (Kurien, 1994).
11. Gas light are started using in 1991 onwards. Attempts were also made by fishermen of Marianad to use battery powered tube lights (PCO, 1995).
12. It later spread to all the 47 coastal villages of Thiruvananathapuram. About 1686 (82 per cent) of plywood boats are using FALs at present (PCO, 1995).
13. It has been pointed out that traditional fishermen operating shore seine used to dump rocks fastened with coconut fronds into sea bottom to attract fish closer to the shore. Fish which got aggregated over the bottom structures were caught by shore seine. This practice was based on their knowledge that fish tend to congregate over bottom structures (Fernandez, 1994).
14. People's Artificial Reefs (PARs) while showing the participation of fishermen, now develop into an important conservation measure of the marine eco-system popularly known as artificial fish habitats. For understanding the formation, evolution and impact of these artificial fish habitats in the Kerala coast see (Fernandez, 1994).

REFERENCES

1. Alagaraja, et al., (1994): "Recent Trends in Marine Fish Products in Kerala with Special Reference to Conservation and Management of Resources", *State Level Seminar,* Kerala Fisheries Society.
2. FAO.,(1974): *Report on a Meeting for Consultations on Under Water Noises*, Report Number 76, Rome.
3. Fernandez, (1994): *Artificial Fish Habitats. A Community Programme for Bio-diversity Conservation,* Fisheries Research Cell, PCO, Thiruvananthapuram.
4. Gillet, P., (1979): *Ten years of Involvement with Fisheries and Fishermen in Kanyakumary District,* KSSS, Publication No. 13, Nagarcoil.
5. Government of Kerala, (1993): "Review of Plan Progress," Fisheries Department Officers Conference.
6. Gulland J.A.,(1979): *The Management of Marine Fisheries,* Scientechnica (Publishers) Ltd, Bristall.
7. Hakkim, A.V.M., (1980): *Mechanisation and Co-operative Organisation: Their Impact on Traditional Fishermen of Kerala,* (Unpublished M.Phil Dissertation), Centre for Development Studies, Thiruvananthapuram.
8. Ibrahim, P., (1986): *Fishing Industry in Kerala: A Study of the Impact of Technological Change,* (Ph.D Dissertation), University of Kerala, Thiruvananthapuram.
9. Kaleeckal J.Jose, et al., (1988): *Oru Samara Katha,* Kerala Swathanthra Matsya Thozhilali Federation, Thiruvananthapuram.
10. Kocheri, T. Fr : (1982): *Need for stopping Trawling in June, July and August,* (Union Report), Kerala Swathanthra Matsya Thozhilali Federation, Kollam.
11. Kottar Social Service Society, (1971): *Mechanisation of Indigenous Fishing Crafts, The Indo - Belgian Fisheries Project on Muttom, A Statistical Study.*
12. Kurien, John, (1994): "Collective Action and Common Property Resources Rejuvenation: The Case of People's Artificial Reefs in Kerala State, India" (Mimeo) Paper presented at the IPFC Symposium on *Artificial Reefs and Fish Aggregation Devices*

as Resource Enhancement and Fisheries Management Tools, Colombo.

13. Kurien , John, and Achari, T.R.T., (1988): "Fisheries Development Policies and and Fisheries Struggle in Kerala". *Social Action,* Vol.38.
14. Kurien and Mathew, (1982): *Technological Change in Fishing: Its Impact on Fishermen,* Centre for Development Studies, Thiruvananthapuram.
15. PCO., (1995): *Problems of Use of Fish Attracting Lanterns in Coastal,* (A Report on the One Day Workshop), October. Thiruvananthapuram
16. Platteau, J.P. et.al, (1985): *Technology and Credit Indebtedness in Marine Fishing- A Case Study on Three Fishing Villages in South Kerala.* Hindustan Publishing Corporation, India.
17. Raj, S.P.J., (1990): "Mapping the Inshore Floor of India", *Sea Food Export Journal,* Vol. 2.
18. Rajan, J.B., (1995): "Fish Attracting Lanterns: Survival Strategy or Market Orientation?" Paper submitted in a workshop on Livelihood *Strategies of Rural Poor and the Environment Challenges Ahead, IIM, Banglore.*

6

Motorisation : The Recovery Phase of Artisanal Fishery

The capitalist intervention in labour process through the medium of technological changes and the inevitable deprivation of the real producers engendered certain unusual changes in the Kerala fishery[1]. One such change is the clout amassed by the marginalised community to intervene in the production process and recapture the control of the labour process [2]. The conflicting interest procreated another phase in the Kerala fishery, popularly known as Motorisation phase[3]. We try to argue in this chapter that the motorisation phase has set in motion certain favourable trends of benefit to artisanal fishermen. The fishermen learned that motorization would strengthen their productive capacity vis-a-vis mechanised sector. The pace of motorisation given in table 6.1 is indicative of this positive expectation.

Table 6.1. Progress of Motorisation Process During 1980s.

Year	*Motorised crafts (Numbers)*	*Percentage increase*	*Cumulative percentage increase*
1981	Marginal	—	—
1982	Marginal	—	—
1983	2,200	—	—
1984	3,965	80	80
1985	6,574	66	146
1987	9,600	46	192

Source :Achari, (1989)

By the close of 1980s only less than five per cent of the artisanal fishermen left behind in the non mechanised category. In this chapter, we are out to probe the viability and sustainability of the new changes and also highlight how the artisanal fishermen succeeded in regaining the lost control of the labour process. In this regard, it is worthwhile to scrutinize the catch data of these two distinct phases of the Kerala fishery- the post-mechanised and post-motorised phases.

The post mechanised period is split into two periods:

1. A period of positive impact, i.e., from 1969 -1974 and
2. The period of negative impact of mechanisation, i.e., from 1975 -1980.

Table 6.2 and 6.3 contains the catch data of these periods respectively.

Table 6.2. Marine Fish Landings of Kerala (1969-74)

Year	*Total Landings (Tonnes)*	*Share of the Mechanised Sector*		*Share of the Artisanal Sector*	
		Quantity	*Percentage*	*Quantity*	*Percentage*
1969	2,94,787	28,177	9.6	2,66,610	90.4
1970	3,92,880	52,771	13.4	3,40,309	86.66
1971	4,45,347	47,291	10.6	3,98,056	89.4
1972	2,95,618	38,648	13.0	2,56,970	86.9
1973	4,48,269	93,659	20.9	3,54,610	79.1
1974	4,20,250	1,01,412	24.1	3,18,845	75.9
Average	3,82,860	60,293	15.3	3,22,567	84.7

Source: PCO, (1991)

These tables (6.2 and 6.3) reveal certain trends in the Kerala fishery. During 1969 - 74 the share of the mechanised sector in the total production has grown on an average of 15.7 per cent, when compared with a share of 9.6 in 1969. It means that the share of the mechanised sector has increased by about one per cent annually. However, the performance of the artisanal sector was dismal. From a 90 per cent share in 1969, its share became 84.7 per cent on an average during 1969-74 (six per cent decline from its 69 level). We would say that year after year, the artisanal sector declined marginally by about one per cent. The average fish landings during this period increased by about 88073 tonnes from that of 1969 level, an increase of about 30 per cent. Thus we see that mechanisation has increased the fish landings tremendously but it has brought in the downswing of the traditional sector. This trend intensified during 1975 - 80.

Table 6.3. Marine Fish Landings of Kerala (1975 -80)

Year	*Total Landings (Tonnes)*	*Share of the Mechanised Sector*		*Share of the Artisanal Sector*	
		Quantity	*Percentage*	*Quantity*	*Percentage*
1975	4,20,836	18,011	42.8	2,40,725	57.2
1976	3,31,047	57,717	17.7	2,72,330	82.3
1977	3,45,037	1,07,424	31.1	2,37,613	68.9
1978	3,75,339	1,17,356	31.5	2,55,768	68.5
1979	3,30,509	94,779	28.7	2,35,768	68.5
1980	2,79,543	1,34,783	48.2	1,44,760	51.8
Average	3,46,719	1,15,362	33.3	2,31,154	66.7

Source: PCO. (1991)

During this period the average fish landings declined by 36141 tonnes when compared with the average production during 1969-75 . This period, generally referred as period of negative impact, however, has not inflicted any negative effect on the out put of the mechanised sector. In fact, the share of this sector in the total production has increased substantially during this period. The average share of the mechanised sector increased to 33.3 per cent from the previous time average of 15per cent. It means that during this period the annual growth in production grew by 29 per cent which was almost double of the previous period. It also means the brunt of the negative growth of the fishery sector was borne by the traditional sector in the form of a drastic fall in the average share of the artisanal sector which declined fi om 84.7 per cent in the previous period to 66.7 per cent. While the annual growth of the artisanal sector declined by one per cent in the previous time period, the declining trend has intensified to three per cent per annum during 1975-80. In short, the impact of mechanisation in the fishery led to a positive effect of increase in out put in the mechanized sector and a negative impact of a continuous fall in the share of the aritisanal sector.

We may now interpret the catch data of the post motorised period (1981-90). Table 6.4 shows that share of the motorised sector has increased tremendously and continuously.

Table 6.4 : Share of the Motorised Sector in the Marine Fish Landings (1981-90)

Year	*Total Landings*	*Share of the Motorised Sector*	
		Quantity	*Percentage*
1981	2,73,978	22,848	8.3
1982	3,25,367	63,050	19.4
1983	3,85,282	99,082	25.7
1984	3,92,895	1,33,319	33.9
1985	3,25,729	1,20,767	37.1
1986	3,83,788	1,86,540	48.7
1987	3,03,286	1,11,208	37.0
1988	4,68,808	2,38,808	50.9
1989	6,47,526	4,06,652	62.8
1990	6,62,890	2,31,547	65.1
Average	4,16,955	1,61,382	38.9

Source: PCO, (1991)

From a little over eight per cent of the share in 1981, the motorised sector has increased its share about two and half times greater on average during 1981-90. However, during this period, the average share of the mechanised sector has also improved but only marginally from 33.3 per cent to 37.3 per cent. This increase in share of the motorised sector was at the expense of the non-motorised sector. Hence, it is essential to assess the impact of motorisation covering the whole artisanal sector. Table 6.5 shows the total fish landings and the respective shares of the mechanised sector and artisanal sector.

The data during 1981 - 92 show that the average share of the artisanal sector was 64.6 per cent. When compared with the average share of 1975-80 period (66.7 per cent) there is only a negligible difference in the share of the artisanal sector between the two periods (2 per cent). However, in terms of the impact of motorisation this figure is of tremendous significance. It points to the potential of the motorisation process to recover the lost grounds of the artisanal fishermen. Even though the aritsanal sector could not improve its share from that of the previous level of 66.7 per cent during 1975-80, the motorisation process succeeded in arresting the continuous deterioration which mechanisation process had inflicted on them.

Table 6.5. Marine Fish Landings of Kerala (1981-92)

Year	*Total Landings (Tonnes)*	*Share of the Mechanised Sector*		*Share of the Artisanal Sector*	
		Quantity	*Percentage*	*Quantity*	*Percentage*
1981	2,73,978	73,056	26.5	22,848	73.3
1982	3,85,367	85,190	26.2	63,050	73.8
1983	3,85,282	98,070	25.5	99,082	74.6
1984	3,92,895	1,29,641	33.0	1,33,319	67.0
1985	3,25,729	1,27,835	39.0	1,20,767	60.8
1986	3,83,788	1,29,526	53.8	1,86,540	66.2
1987	3,03,286	1,51,178	49.9	1,12,208	50.2
1988	4,68,808	1,96,780	42.0	2,38,808	58.0
1989	6,47,526	2,08,013	32.1	4,06,652	67.9
1990	6,62,890	4,31,343	34.9	2,31,547	65.1
1991	5,64,161	2,19,684	38.9	3,44,477	61.0
1992	5,60,742	2,55,138	45.5	3,05,604	54.5
Average	4,46,204	1,75,455	37.3	1,88,742	64.6

Source: PCO. 1991 and Achari, (1994)

Further, during the motorisation phase the artisanal fishermen succeeded in keeping at bay the ruthless deprivation of output and income done by mechanised fishing. The average catch in the mechanised phase (1969-80) was 364789 tonnes and in the motorised phase (1981-92) was 446204 tonnes. Thus in the motorised phase, average catch has increased by more than 22 per cent. This shows that the artisanal sector succeeded in augmenting the harvesting capacity of the artisanal fishermen and thus counter balance the deprivation meted out to them.

A disaggregated view of the catch data in the post-motorised period also reveals that the quantum of harvest of artisanal sector has increased at rates higher than that of total production. For instance, when total production has increased continuously during 1981-84, the harvest of the artisanal sector also increased continuously and substantially. Similar is the trend during 1988-90.

Thus aggregated and disaggragated views of the catch data of the post-motorised period reveal that the motorisation process enabled the artisanal fisheries to augment their harvesting capacity, enabling them to make a recovery from the deprivation caused by capitalist intrusion in their traditional domain[4]. However, it is our endeavour to probe further whether the new technology is economically viable, the changes are sustainable and whether it would enable them to control the labour process to give them emancipation from capitalist dominance. This task is accomplished by analysing the primary data collected from important centres of motorised fishing all over the Kerala coast. It may be cautioned that the adoption of the new technical alignment is not uniform through out the state and even in the centres where there had some uniformity, the competition among the fishermen added newer dimensions to the motorisation process. For these reasons, we do the analysis in terms of two regional classifications, viz., North, and South Zones.

The North Zone comprises of the districts of Kasargod, Kannur, Kozhikod and Malappuram. The South region includes Alappuzha, Kollam and Thiruvananthapuam districts. In the North Zone, the dominant craft-gear combination during the pe-motorised period was the dug-out-Koruvala/Kollivala units. Since motorisation, drastic changes occurred in the fishing techniques at varying degrees all over this region. However, the operation of a new type of fishing technique called ring-seine came up fast. With the spread of this new techniques, new combinations emerged in the different districts of the North Zone.

In Kasargod, the ring seine known as Rani Vala was originally operated by two boats[5]. One boat while holding one end of the net stable, the other boat encircles the shoal. However, later it evolved into a four boat operation where the additional crafts participated as holders of float lines and as carriers. All crafts involved in the ring seine operation used 8 H.P or 15 H.P engines in their boats.

In other districts of the North Zone also, the Ring seine technology replaced the old Kollivala units. The fishermen organized ring seine units by pooling their existing crafts and gears. Thus by making some marginal changes in their equipments and pooling their resources, they succeeded in incorporating the new technology with out much economic hardship. Based on the number of fishermen who joined to form ring seine units, the size of the ring seine varied from two to eight crafts operation. However, four craft operation became increasingly popular.

Two novel versions of ring seine combination in the North Zone were plank boat and plywood boat versions. It is mainly the influence of the Southern fishermen which made northerners to adopt plank boat ring seine combination. Unique advantages associated with plywood boats made others to adopt the plywood ring seine combination. However, fishermen increasingly used plywood ring seine combination and thus enabling plywood ring seine combination became the dominant craft gear combination in the Northern Zone.

Besides the ring seine operation the fishermen in this Zone resort to gillnet, mini trawl, and hook and line fishing. It is the medium crafts which are generally used for gillnet and mini trawl operations. Since very few fishermen use hook and line it is not given importance in the study.

A systematic account of the motorised crafts is not available in the State. However, a census of the artisanal marine fishing fleet of Kerala conducted by South Indian Federation of Fishermen Societies (SIFFS) in 1991 provide, the best data of the motorised crafts and gears. With this basic information, the study proceeded by collecting sample data from across the beaches covering important fishing centres of Kerala. Generally the study intends to collect five per cent sample from each category of craft gear combinations at random.

In the North Zone 681 crafts used ring seine. Of this 347 were plank built, 282 were dug out crafts and 52 were made of plywood. While five per cent of dug outs and plank ring seines are taken as the samples from the beaches of important fishing centres, in the plywood ring seines 10 units are taken since it is fast emerging as an important category.

The gillnet and mini trawl operations are generally conducted in medium crafts. Even though information about the total medium crafts in the North Zone is available, all these crafts are not participating in gill net fishing.

The SIFFS study has accounted the distribution of the crafts in terms of gears in different districts. Accordingly, in the Northern Zone, about 383 crafts were aligned exclusively with large gill nets and about 1037 crafts were aligned with small gillnets. From this population in the North Zone five per cent sample has been taken to analyse the gill nets, as large gillnet and small gillnet operations. Thus taking the entire motorised fishing scenario in the North Zone, a five tier classification of the fishing units are made viz;

1 dug-out Ranivala (Ring Seine) units
2 plank Ranivala units
3. ply-wood Ranivala units
4. gillnet units (Small)
5. gillnet units (Large)

In the South region three district are covered, viz., Alappuzha, Kollam and Thiruvananthapuram. In Alappuzha and Kollam districts, ring seine operations were in large scale done along with gill netting and sparingly hook and lines. However, in Thiruvananthapuram district ring seine operations are completely absent; gillnetting and hook and lines were the major fishing operations. Even though the SIFFS study had made one of the classifications of fishing crafts as plank transom seen both in Alappuzha and Thiruvananthapuram, it was found that only in Alappuzha such crafts exists at present and it was totally extinct in Thiruvananthapuram. Analogous to the North, a five tier categorization is adopted in the South. These are:

1. plank-Ring seine units (Very large)
2. plank-Ring seine units (Large)
3. plank-Transom units
4. gillnet units (Large & medium)
5. plywood Gillnet units

The total number of plank ring seine units in the South region is 556. A sample of 28 units were taken from major fishing centres at random. In the large plank ring seine category, there were 256 fishing units and a sample of 13 units were taken by the same procedure. In the plank transom type, there are 640 units in Alappuzha. A sample of 32 units were taken to represent the plank transom units. In the large and medium gillnet category in the South region as a whole there were 393 crafts, of which 20 units were taken to represent this population. In the plywood gill net category mainly both medium and small nets, there were 1560 units in the whole region and a sample of 78 units were taken to represent this group.

The purpose of the Primary Data Analysis were

1. to inquire into the economic conditions of the motorised fishing units,
2. whether the new technology inducted provided a profitable venture for the fishermen, and,

3. whether the new changes provided a control over the labour process in fishing to the traditional fishermen which had lost to them due to the capitalist intrusion.

Cost and Return Analysis of Motorised Fishing Units

We make a cost and return analysis of motorised fishing units, initially technology wise and then region wise. In the five fold classification of fishing units in the North Zone, the sample of the dug-out ring seine units reveal that all units make economic gains from their fishing activity. A comparison of cost and return measured as profit shows that all the sample units in this category make profit from their activity (See table 6.6). Table 6.6 shows the distribution of all the fishing crafts of the North Zone according to their economic performance in terms of profit or loss or as marginal survivors.

Table 6.6 points out that while more than 90 per cent of plank-ring seine and gillnet (small) units make profits, 20 per cent of the plywood ring seine and four per cent of gillnet units (small) incur loss. The marginal units are totally absent in ring seine category where as in the gillnet category a little more than five per cent are marginal survivors.

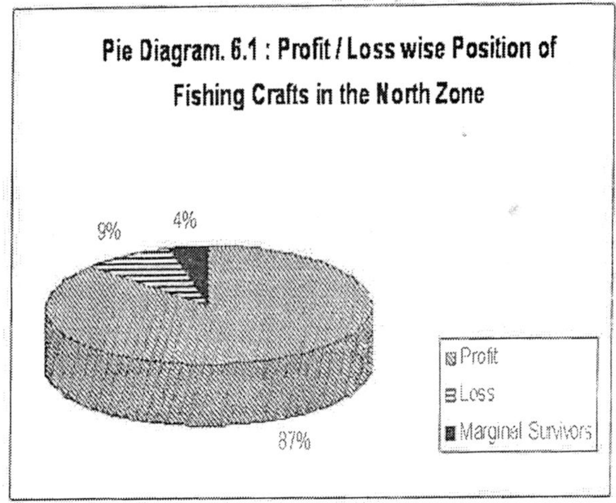

While 87 per cent of the fishing units make economic gain, nine per cent incur loss and four per cent are marginal survivors. Even though the percentage of profit making fishing unit is a desirable piece of information, the economic condition of the motorised sector is not fully revealed. Besides the number of profit making units, one would also know the level

Table 6.6 : Distribution of Fishing Units According to Profit / Loss or Marginal Units in the North Zone.

Major fishing crafts	*Dugout ring seine units*		*Plank ring seine units*		*Plywood ring seine units*		*Gillnet units (large)*		*Gillnet units Small)*		*All units*	
	Number	*Percentage*	*Number*	*Percentage*	*Number*	*Percentage*	*Number*	*Percentage*	*Number*	*Percentage*	*Number*	*Percentage*
Profit	17	100.00	14	93.33	8	80.00	13	68.42	47	90.38	99	87.61
Marginal units	-	-	-	-	-	-	1	5.26	3	5.77	4	3.54
Loss	-	-	1	6.67	2	20.00	5	26.32	2	3.85	10	8.85
Total	17	100.00	15	100.00	10	100.00	19	100.00	52	100.00	113	100.00

Source : Survey dataThe economic position of the fishing units in the North region as a whole is shown in the Pie Diagram. 6.1

of profit earned. Technology wise distribution of the crafts over different levels of profit is shown in table 6.7.

Table 6.7 reveals that in the dugout-ring seine category 70 per cent of the crafts earn more than 50 per cent profit on their investment. The next type which gets a substantial return on their investment belongs to the gillnet (small) units. Fifty per cent of the crafts in this category earn a profit of more than 50 per cent on their investment. The crafts under plank ring seine and plywood ring seine also earn a fair level of profit. More than 66 per cent units in the former category and 50 per cent in the latter earn a profit level between 11 per cent and 50 per cent. In the case of the remaining large gillnet units, while 32 per cent of the crafts incur loss, 11 per cent earn a profit of more than 50 per cent and 37 per cent of the units get a return in the range of 10-50 per cent on their investment.

It is also interesting to understand the comparative performance of these different technical alignments in terms profit. For this we may compute the average level of profit in each category. This is given in table 6.8.

Table 6.8 shows that the dugout - ranivala combination reap the highest average return of 83 per cent followed by small gillnet units, of about 56 per cent. Even though, the other types of fishing units also make profit, there is considerable difference in the profit rate. Plank-ranivala and, plywood ring seine are basically the same type of technology except for the differences in crafts. However, with regard to the level of profit realised, these units are different when compared with dug-out rani vala combination, the difference between these two groups being 67 per cent. In the case of gillnet operations also, the picture Is almost same. The large gill nets secure only less than one fourth of the profit rate earned by small gill net units.

There are certain clear reasons for such wide differences in the level of profit. In the case of plank and ply-wood ring seine units, the investments are substantially high when compared with dug-out ranivala combination. The investment in plank-ranivala is more than 25 per cent higher to that of dugout-ranivala and in the case of plywood-ranivala, the investment is more than 39 percent higher. In the case of operating expenses also, the trend is same but more pronounced. In the case of plank-ranivala and plywood-ranivala, the operating expenses are higher by about 94 per cent and 101 per cent respectively when compared with that of dug-out ring seine combination. Higher investment and higher running expenditures have not brought forth any higher returns on these fishing units. On the

Table 6.7 : Distribution of Fishing Units According to Percentage Level of Profit in North Zone

Range of profit (Percentage)	*Dugout ring seine units*		*Plank ring seine units*		*Plywood ring seine units*		*Gillnet units (large)*		*Gillnet units Small)*		*All units*	
	Number	*Percentage*	*Number*	*Percentage*	*Number*	*Percentage*	*Number*	*Percentage*	*Number*	*Percentage*	*Number*	*Percentage*
1-10	3	17.65	4	26.67	3	30.00	3	15.79	4	7.69	17	15.04
11-25	-	-	5	33.33	3	30.00	5	26.32	6	11.54	19	16.81
26-50	2	11.76	3	20.00	2	20.00	2	10.53	11	21.15	20	17.70
51-100	3	17.65	2	13,33	-	-	2	10.53	15	28.85	22	19.47
101-150	7	41.18	-	-	-	-	-	-	5	9.62	12	10.62
151-200	2	11.76	-	-		-	-	-	5	9.62	7	6.19
> 200	-	-	-	-	-	-	-	-	1	1.92	1	0.89
No. Profit	-	-	-	-	-	-	1	5.25	3	5.77	4	3.54
Loss	-	-	1	6.67	2	20.00	6	31.58	2	3.84	11	9.74
Total	17	100.00	15	100.00	10	100.00	19	100.00	52	100.00	113	100.00

Source : Survey data

Table 6.8: Average Annual Cost and Return of the Major Fishing Units in the North Zone.

Sl. No	*Items (Rs. In lakhs)*	*Dug out ring seine units*	*Plank-ring seine units*	*Plywood ring sience units*	*Gillnet units (Large)*	*Gillnet units (Small)*
1.	Capital investment	7.34	9.20	10.22	1.47	0.78
2.	Operating expenses	3.52	6.82	7.09	0.82	0.74
3.	Receipt	26.07	17.75	17.27	2.06	2.28
4	Sales Commission	0.78	0.88	0.86	0.08	0.10
5.	Gross income	21.77	10.04	9.32	1.16	1.43
6.	Crew remuneration	14.00	6.02	5.85	0.59	0.82
7.	Net income before depreciation	7.77	4.02	3.46	0.52	0.61
8.	Net income after capital depreciation allowance (at 10 per cent rate)	7.04	3.09	2.44	0.37	0.53
9.	Net income after interest charges (at 12 per cent rate)	6.08	1.99	1.21	0.19	0.44
10.	Rate of return on capital	83	22	12	13	56

Source : Survey data

contrary, the sample data shows that returns on the dug-out ranivala combination is higher than the other similar units. Even though the exact reasons for higher returns by the dug-out units are unknown, what is obvious is that these dugouts are indigenous while others are imported from the southern region. This might have accounted for some differences in productivity. In the case of gill net units also, the larger units incur higher capital investment and running expenses but have not resulted in any proportionate increase in the receipts.

We may now focus on the economic conditions of the major fishing units in the South Zone. Likewise in the Northern fishery, a technology wise distribution of fishing crafts according to loss or profit condition is shown in table 6.9.

Table 6.9 shows that in the case of small gillnet units some fishing units are incurring loss (17.24 %). Two per cent in this category are marginal survivor's while more than 80 per cent of the crafts are profit making units. In all other types of technical alignment 100 per cent of the crafts are profit earners.

The distribution of the fishing crafts in accordance with the level of profits earned is shown in table 6.10.

It reveals that more than fifty per cent of both the very large and large fishing units of the plank-ring seine category make a profit level of 50 per cent or more. In the case of large gill net units all crafts are earning a profit level of more than 25 per cent. More than 80 per cent of the plank transom units earn profit level of 25 per cent or more. As against these trends in the case of small gill net units about 17 per cent of the units are incurring loss and 78 per cent of the units are getting a profit ranging between 1-25 per cent. A lower profit level for majority of fishing units in the small gill net category and the loss suffered by a considerable section among them shows that the units in the small gill net category is exposed to severe competition for fishing space resulting in lower catches.

A comparative analysis of major technical alignments in the southern fishery is attempted in terms of average investments, and returns in table 6.11.

A study of the table 6.11 reveals that there is no direct relation between capital investment and profit. Higher investment in capture fishery need not fetch higher returns. However, table 6.11 shows that units with higher investment fetches higher receipts. But higher receipts do not turn into higher returns because it also entails higher operational expenses.

Table 6.9 : Distribution of Fishing Units According to Profit / Loss or Marginal Units in South Zone

Types of crafts	*Plank ring seine units (very large*		*Plank ring seine units (large)*		*Plank transom units*		*Gillnet units (large)*		*Gillnet units Small)*		*All units*	
	Number	*Percentage*	*Number*	*Percentage*	*Number*	*Percentage*	*Number*	*Percentage*	*Number*	*Percentage*	*Number*	*Percentage*
Profit	28	100.00	13	100.00	31	100.00	21	100.00	70	80.46	163	90.56
No Profit	-	-	-	-	-	-	-	-	2	2.30	2	1.11
Loss	-	-	-	-	-	-	-	-	15	17.24	15	8.33
Total	28	100.00	13	100.00	31	100.00	21	100.00	87	100.00	180	100.00

Source : Survey data

Table 6.10 : Distribution of Fishing Units According to the Level of Profits in South Zone.

Range of profit (Percentage)	*Dugout ring seine units*		*Plank ring seine units*		*Plywood ring seine units*		*Gillnet units (large)*		*Gillnet units Small)*		*All units*	
	Number	*Percentage*	*Number*	*Percentage*	*Number*	*Percentage*	*Number*	*Percentage*	*Number*	*Percentage*	*Number*	*Percentage*
1-10	1	3.57	-	-	1	3.23	-	-	27	37.03	29	16.11
11-25	6	21.43	1	7.69	5	16.13	-	-	39	44.83	51	28.33
26-50	7	25.00	4	30.77	11	35.48	4	19.05	4	4.60	30	16.67
51-100	13	46/43	3	23.08	9	29.03	2	9.52	-	-	27	15.00
101-150	1	3.57	3	23.08	4	12.90	10	47.62	-	-	18	10.00
151-200	-	-	2	15.38	1	3.23	4	19.05	-	-	7	3.89
>200	-	-	-	-	-	-	1	4.76	-	-	1	0.56
No. Profit	-	-	-	-	-	-	-	-	2	2.30	2	1.11
Loss	-	-	-	-	-	-	-	-	15	17.24	15	8.33
Total	28	100.00	13	100.00	31	100.00	21	100.00	87	100.00	180	100.00

Source : Survey data

Table 6.11 : Average Annual Cost and Return of the Major Fishing Units in the South Zone

Sl. No	*Items (Rs. In lakhs)*	*Dug out ring seine units*	*Plank-ring seine units*	*Plywood ring sience units*	*Gillnet units (Large)*	*Gillnet units (Small)*
1.	Capital investment	5.91	4.53	0.87	1.27	1.93
2.	Operating expenses	7.23	3.18	0.89	1.11	1.13
3.	Receipt	18.62	17.10	2.30	5.77	2.70
4	Commission	0.93	0.85	0.12	0.28	0.12
5.	Income	10.45	13.06	1.29	4.38	1.45
6.	Crew remuneration	6.26	7.84	0.65	2.63	0.90
7.	Net income before depreciation	4.17	5.23	0.65	1.75	0.55
8.	Net income after capital depreciation (at 10 per cent rate)	3.58	4.77	0.56	1.62	0.36
9.	Net income after interest charges (at 12 per cent rate)	2.88	4.23	0.46	1.47	0.13
10.	Rate of return on capital (percentage of profit)	49	93	53	116	7

Source: Survey data

Ratios of operating expenses to average receipts of these technical alignments is shown in table 6.12.

Table 6.12. Ratios of Operating Expenses to Receipts of the Major Technical Alignments in the South Zone

SL No.	*Technical*	*Plank ring seine units (very large)*	*Plank ring seine units large)*	*Plank transom m units*	*Gillnet) units (large)*	*Gillnet units (small)*
1	Operating expenses (in Rs.)	202.51	41.28	27.67	22.27	98.47
2	Receipts (in Rs.)	521.25	222.26	71.37	115.37	235.33
3	Ratio of operating expenses to receipts	0.39	0.19	0.39	0.19	0.42

Source: Survey data

Information of table 6.2 vis-a-vis that of table 6.11 point out that though higher receipts emerge from higher capital investment, profit level is determined by operational expenses. For those units with lower operational expenses give higher returns. Thus in the southern fishing large gill net units and large plank ring seine units produce higher rate of return on capital than that of other units.

It is also interesting to make a comparative study of the two regions. The bar diagram 6.1 makes such an attempt. The bar diagram shows the comparative values of the average cost and return of the fishing operations of the major fishing crafts in both regions.

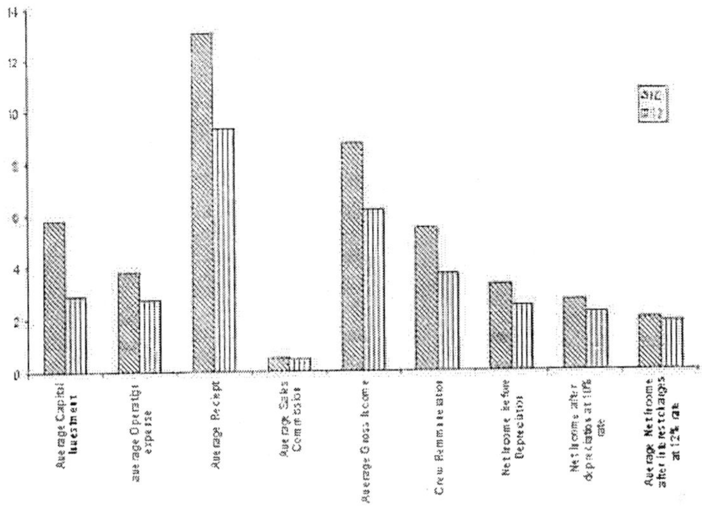

BAR DIAGRAM 6.2

COMPARATIVE VALUES OF THE AVERAGE COST AND RETURN OF MAJOR FISHING CRAFTS OF NORTH AND SOUTH ZONE

A comparison of the bar chart shows that the capital investment in the North Zone is double higher than that of the South Zone. Similarly, the operating expenses in the Northern part is 40 per cent higher than the South. The sales commission given by the fishermen in the North is higher by 17 per cent to that of the fishermen of the South. The crew remuneration is also higher in the North by about fifty per cent. Even though, there is substantial difference in the average receipts in two regions (40 per cent higher than in the North), the higher costs of fishing activities in the North overrides the higher returns and results in lower return on capital in the North. The average rate of return in the South is 64 per cent whereas in the North it is 37 per cent thus resulting the rate of return 27 per cent less in the North compared with the South.

The cost and return analysis of the Kerala fisheries shows that motorisation has provided fishermen a profitable economic activity. However, certain physical indices would throw more light into the economic efficiency of the motorisation. Two such physical indices are the catch per unit effort and catch per unit energy. The former is a measure of catch divided by man hours while the latter is a measure of catch divided by energy expenses.

Table 6.13 and 6.14 shows per day measure of the total catch, man days and energy expenses for the whole sample in both Zones. We derive the catch per unit effort and catch per unit energy from these figures.

Table. 6.13. Technology wise Measure of Total Catch, Energy and Mandays of North Zone

SL No.	*Description of technical alignments*	*Catch (in Kg.)*	*Energy expenditure In Rs.)*	*Mandays*	*Catch per unit effort (in Kg.)*	*Catch per unit energy in Kg.)*
1.	Dugout ring seine units	21,100	41,205	5,595	3.77	0 51
2.	Plank ring seine units	13,470	59,780	3,930	3.43	0 23
3.	Plywood ring seine units	7,300	41,675	5,600	1.30	0 18
4.	Gillnet units (large)	666	8,345	538	1.24	0 08
5.	Gillnet units (small)	5,280	19,055	991	5 33	0.28

Source: Survey data

Table . 6.14. Technology wise Measure of Total Catch, Energy and Mandays of South Zone

SL No.	*Description of technical alignments*	*Catch (in Kg.)*	*Energy expenditure In Rs.)*	*Mandays*	*Catch per unit effort (in Kg.)*	*Catch per unit energy in Kg.)*
1.	Plank ring seine units (very large)	19,815	76,730	7,300	2.71	0 26
2.	Plank ring seine units (large)	6,420	16,715	1,805	3.56	0.38
3	Plank transom units	593	11,195	740	0.80	0 05
4.	Gillnet units (large)	3,700	8,870	484	7 64	0 42
5.	Gillnet units (small)	6,020	41,965	2,488	2 42	0 14

Source: Survey data

The catch per unit effort and energy of the different technologies derived in table 6.13 and 6.14, show that in both Zones, there are substantial differences in these two measures. In the North Zone the catch unit per effort varies between 1.24 kg and 5.33 kg implying a more than three times difference in productivity among the different technical combinations. The variation in the catch per unit energy is 0.43 kg, the highest being 0.51 kg and the lowest 0.08 kg, a difference of more than five times. These show that the productivity difference in terms of fuel efficiency is greater than in terms of catch. At specific level in the North Zone, catch per unit effort is the highest for gillnet (small) units among all the crafts, followed by dug-out ring seine units. In the matter of catch per unit energy the order is reversed. Dugouts are the traditional crafts with which Northern fishermen is long associated and this might have provided some advantages culmunating in higher catch per unit effort and catch per unit energy. While the plank ring seine units are also very close to dugouts in catch per unit effort, Plywood ring seines have only about one-third of the catch per unit effort of dugouts.

In the South region also the catch per unit effort and energy show wide difference. In the catch per unit effort the difference is 6.84 kg (more than eight and half times), while in the catch per unit energy this is 0.37 kg (about seven and half times)

In the Southern fishery, it is large and medium gillnet fishing units that perform well in terms of catch per unit effort and catch per unit energy. One important reason for this is that the energy expenses and man hours spend in fishing are substantially less in this particular technical alignment. Most of these fishing units fish between 20-25 km range in the sea where as all other craft type except plank transom units goes well beyond 35 km. This reduces the fuel costs and man hours spent. Besides most of these units possess different types of gears to provide flexibility in fishing depending upon the availability of different varieties of species. The result was that catches are relatively high when compared with other types of fishing units. In the case of productivity indices plank-transom units are abysmally low both in catch per unit effort and catch per unit Energy. The plank transom units specialize in trawling by using mini trawl nets. Since they capture prawns primarily, the relatively lower availability of prawns cause those units faring little against other fishing units. It is mainly the lower quantity of catches that results in poor catch per unit effort and energy.

A comparison of differences in productivity between North and South Zones reveal that the productivity differences in terms of catch and fuel efficiency are higher in the South than in the North. However, the inter technology difference between these indices show that its is much wider in the North than in the South.

Even though, the technology wise inter and intra comparison of the productivity indices entail substantial differences, at an aggregate level, the catch per unit effort and catch per unit energy are almost equal in both Zones. This is evident from table 6.15.

Table 6.15. Catch Per Unit Effort and Energy in the North and South Zones

Sl. No.	*Zones (in Kg.)*	*Catch expenditure (in Rs.)*	*Energy*	*Mandays*	*Catch per unit effort (in Kg.)*	*Catch per unit energy (in Kg.)*
1.	North Zone	47,816	1,70,060	16,654	2.87	0.28
2.	South Zone	36,548	1,55,475	12,817	2.85	0.24

Source: Survey data

Almost equal values of productivity indices in both zones indicate a fairly distributed coastal specie along south west part of coastal Kerala. While different technical components are developed in the artisanal sector since 1980s, the equality of energy index highlights that the energy component of these diversified fishing techniques are also generally the same.

Finally, we may look at the catch per unit effort and energy for the whole fishing units in Kerala. This is shown in table 6.16.

Table 6.16 : Catch Per Unit Effort and Energy of Fishing Units (All Kerala)

Sl. No.	*Zones (in Kg.)*	*Catch expenditure (in Rs.)*	*Energy*	*Mandays*	*Catch per unit effort (in Kg.)*	*Catch per unit energy (in Kg.)*
All Kerala		84,364	3,25,535	29,471	2.86	0.26

Source: Survey data

It is worthwhile to compare the index of catch per unit effort with that of earlier measures at different time periods. Such a comparison is made in table 6.17.

Table 6.17 : Comparison of Catch Per Unit Effort at Different Time Periods

Year	*Catch per unit effort (in Kg.)*
1980-81[1]	3.20
1988-89[2]	3.85
1996-97	2.86

Source: 1 & 2 PCO and SIFFS study, (1991), and the rest Survey data

Table 6.17 shows that the motorised sector succeeded in enhancing productivity from 3.20 kg in 1980-81 to 3.85 kg during 1988-89. This has happened at the height of motorisation process. However, it decreased to 2.86 kg in 1996-97. This fall indicates the over investment occurred in the motorized sector.

NOTES

1. Theoretical explanation of this unusual change is discussed in chapter two. The labour process changes have occurred in a typical way because of the peculiarities of the socio-economic conditions in LDCs on the one hand and the dependency relations of such economies on the other.
2. These forces have been termed as 'constructive responses' in the previous chapter (Chapter five). Among such forces, the technological responses in the from of innovation/adaptation is the dominant one.
3. Attempts of motorisation of the country crafts by individuals and collective efforts of fishermen and the relatively successful completion of such efforts in to motorisation phase show that technology is not neutral. In fact, technology has a class character.
4. It has been pointed out that since 1990 the share of the mechanised sector and motorised sector has stabilised at one third and two third of the total production respectively (Govt. of Kerala, 1993). It implies that the motorsied sector has made a permanent retrieval from capitalists onslaught. Given the fact that the mechanised boats were mainly owned by capitalists, their loss of control over the capitalist production of fishery will induce them to shift strategically their capital elsewhere.
5. The origin, development, operation and other related aspects of ring seines are discussed by J.B. Rajan (1993).

REFERENCES

1. Achari, T.R.T., (1989), "Growth of Motorised Crafts in 1980s", Paper Presented in National Workshop on *Technology for Small Scale Fish Workers,* Trivandrum.

2. Govt.of Kerala, (1993), *Motorisation of Fishing Units: Benefits and Burdens,* Report of the Study Techno -Economic Analysis of Motorisation of Fishing Units Along the Lower South-West Coast of India, PCO, SIFFS.

3. PCO and SIFFS, (1991), *Fisheries Development and Management Policy.* Rajan,J.B.,(1993), *Techno -Economic Study of Ring Seine Fishing in Kerala,* PCO, Thiruvananthapuram.

7

Reconstitution of Labour Process : Emerging Trends

The changes in labour process in the context of conflicting relation between the capitalist and the deprived fishermen paved the way for the formation of a new phase in Kerala fishery - the motorisation phase. The new phase which itself is beset with problems, however, has resulted in stabilising the artisanal share in production at 2/3rd level. The aritisanal fishermen succeed at this mainly because the new technology augmented their harvesting capacity. While the motorization process has accorded economic gains to the fishermen, more striking is the phenomenon of creating an environment, where the artisnal sector potentially challenged the capitalist intrusion and gradually began to regain the lost control of labour process. In this chapter we delineate the specifics of the labour process changes which have empowered the traditional fishermen.

The advent of an alternative technology by individual and collective efforts of fishermen has provided a technical base to challenge capitalist technology. The know how of making plywood-boats, the skill of conversion of the country crafts to fix OBMs, the development of more efficient nets, new active capturing techniques and moreover a "flexible technical package" suited to local conditions made the fishing efforts more productive. This has enhanced their productive capacity and achieved the empowerment of fishermen. At the outset, it has improved their sea going capacity. Table 7.1 shows the sea faring capacity of fishing units in the artisanal sector in the post-motorisation phase.

Table 7.1 shows that about 55 per cent of the crafts are able to fish more than 30 meters depth or more. The fact that majority of the crafts are capable of reaching 30 meters and beyond is an indicator of improved sea

Table 7.1 : Sea Faring Capacity of Major Fishing Units Since Motorisation.

Major fishing Crafts (Depth in (Meters	*Dugout ring seine units*		*Plank ring units*		*Plywood ring seine units*		*Plank transom units*		*Gillnet units (large)*		*Gillnet units Small)*		*All units*	
	Number	*Percentage*	*Number*	*Percentage*	*Number*	*Percentage*	*Number*	*Percentage*	*Number*	*Percentage*	*Number*	*Percentage*	*Number*	*Percentage*
Up to 25	12	70.59	12	21.05	-	-	31	100.00	20	51.28	19	13.67	94	32.08
Up to 30	3	17.65	6	10.53	5	50.00	-	-	4	10.26	21	15.11	39	13.31
Up to 35	1	5.88	13	22.81	3	30.00	-	-	7	17.95	35	25.17	59	20.14
Up to 40	1	5.88	23	40.35	2	20.00	-	-	6	15.38	26	18.71	58	19.80
> 40	-	-	3	5.26	-	-	-	-	2	5.13	38	27.34	43	14.67
Total	17	100.00	57	100.00	10	100.00	31	100.00	39	100.00	139	100.00	293	100.00

Source : Survey Data

faring capacity of traditional fishermen[1]. In the ring seine category while the average depth of operation during 1988-89 was 23 meters, the study shows more than 50 per cent of the crafts in this category operate 30 meters depth or more at present. In the gillnet category (all types of gillnet together) 64 per cent of the crafts are operating 30 meters or more. The average depth of the major fishing crafts could operate at present with that of 1988-89 is shown in figure 7.1.

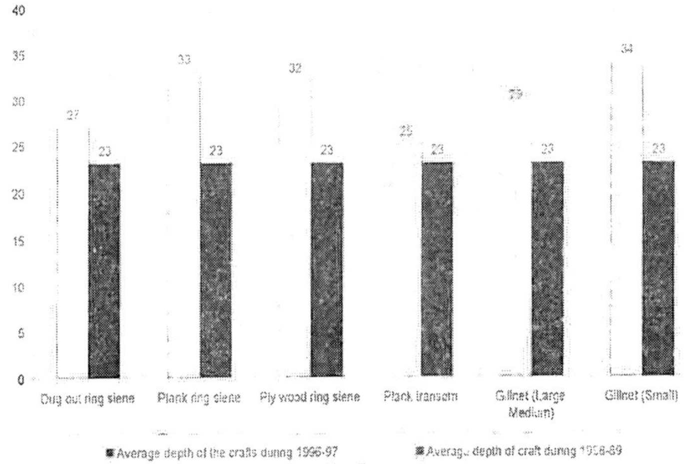

Fig. 7.1 Average Fishing Depth of Major Crafts (in meters)

Source: Average depth of the crafts during 1996-97 - Survey data. Average depth of the crafts during 1988-89 - SIFFS study, (1991)

The extent of the increase in sea faring capability can be learned by comparing non-motorised and motorised sector on the one hand and motorised sector at different periods on the other. Figure 7.2 make such a comparison.

The fishing capacity of the traditional sector was further strengthened by their ability to lengthen their fishing time through motorisation. Table 7.2 shows how long different crafts scan engage in fishing.

Table 7.2 reveals that about 71 per cent of the crafts engage in fishing activity for about 6 - 8 hours, while another 21 per cent spend 9-11 hours and more. Technology wise, it is the ring-seine units that spend longer hours in fishing than the crafts in the gillnet category. Thirty per cent of the ring seine crafts spend 9-11 hours to fish while in the gillnet category 15 per cent of the crafts are capable of fishing the same time. Among the ring seine crafts plywood ring seines are capable of engaging more time in fishing than the other two types. While 40 per cent in plywood ring seines work for about 9-11 hours, it is only 21 per cent which could work for same hours in the dug out and plank ring seine category. This efficiency of plywood boats to engage more time in fishing stems from the versatality associated with the plywood boats. A comparison of the average fishing timings of the crafts with that of 1988-89 and that of the non motorised sector shows that (figure 7.3) the fishing time of motorised crafts have increased considerably.

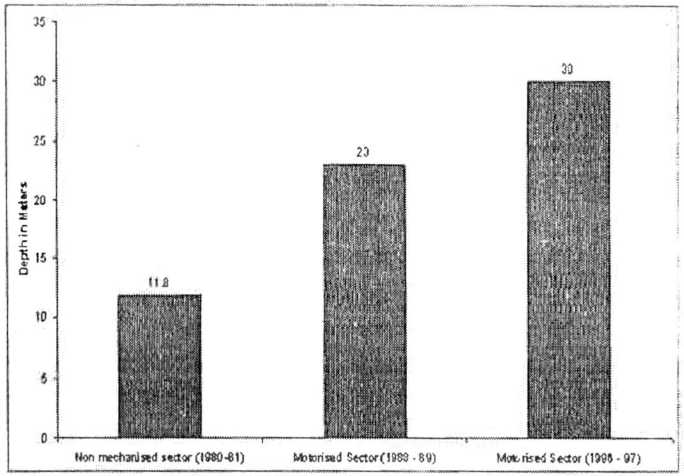

Fig. 7.2 Sea faring capacity of Artisanal Sector at Different Time Periods.

Source:	NM *Sector (1980-81)*	*SIFFS Study*
	M. Sector (1988-89)	"
	M. Sector (1996-97)	Survey data

The new strength to remain in sea for longer hours in fishing enabled them to cast their nets repeatedly either to have a fabulous catch or to cover a missed attempt, both augment productive capacity.

With the new technology, they are not only capable of fishing longer time but could fish larger days too. Their enhanced capacity to fish larger days in shown in table 7.3.

Table 7.2 : Crafts According to Fishing Time

time spend In the Crafts Sea houts	Dugout ring seine units		Plank ring units		Plywood ring seine units		Plank transom units		Gillnet units (large)		Gillnet units Small)		All units	
	Number	Percentage	Number	Percentage	Number	Percentage	Number	Percentage	Number	Percentage	Number	Percentage	Number	Percentage
<5	1	5.88	2	3.51	-	-	-	-	10	25.64	14	10.07	27	9.21
6-8	8	47.06	42	73.68	6	60.00	25	80.65	18	46.15	108	77.70	207	70.65
9-11	5	29.41	13	22.81	4	40.00	6	19.35	9	23.08	16	11.51	53	18.09
>12	3	17.65	-	-	-	-	-	-	2	5.13	1	0.72	6	2.05
Total	17	100.00	57	100.00	10	100.00	31	100.00	39	100.00	139	100.00	293	100.00

Source : Survey data

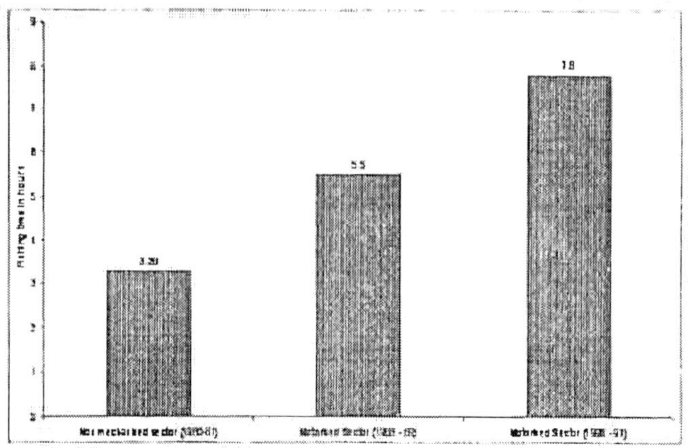

Fig. 7.3. Average Fishing Time of Crafts at Different Time Periods.

Source : N.M Sector (1980-81)	SIFFS Study (1991)
M.Sector (1988-89)	"
M.Sector (1996-97)	Survey data

Table 7.3 shows that more than 63 per cent of the fishing units are capable of fishing 200 days or more. Crafts with less than 150 fishing days has come down to just 10 per cent. A technology wise comparison of productivity of crafts in terms of fishing days shows the gillnet units are ahead of ring seine units. While 94 per cent of gillnet units fish more than 150 days the corresponding figure in the ring seine category is 77 per cent. Thus we see that motorisation has enabled the fishermen to increase their productive capacity at different levels. The new technical alignment gave them the power to obviate, to some extent, the adversities of the weather conditions, enabling them to fish longer days. The new technology provided them the power of control over speed enabling them to stay longer hours further deep into the sea.

The new technology has brought in empowerment of fishermen by offering scope for re-organisation of traditional sector. New changes have provided opportunities for division of fishing activity into different technical alignments (different craft gear combinations and fishing techniques) providing fishermen the facility of specialising in a particular craft-gear or harvesting-technology. The quality of the reconstituted labour process is to be gauged from the following:

Table 7.3: Crafts According to Fishing Days

Fishing days hours	Dugout ring seine units		Plank ring units		Plywood ring seine units		Plank transom units		Gillnet units (large)		Gillnet units Small)		All units	
	Number	Percentage	Number	Percentage	Number	Percentage	Number	Percentage	Number	Percentage	Number	Percentage	Number	Percentage
< 100	3	17.65	-	-	-	-	-	-	-	-	-	-	3	1.03
101-150	6	35.29	6	10.53	4	40.00	-	-	5	12.82	5	3.60	26	8.87
151-200	8	47.06	11	19.30	4	40.00	2	6.45	7	17.95	47	33.81	79	26.96
201-250	-	-	10	17.54	2	20.00	20	64.52	23	58.97	36	25.90	91	31.06
>250	-	-	30	52.63	-	-	9	29.03	4	10.26	51	36.69	94	32.08
Total	17	100.00	57	100.00	10	100.00	31	100.00	39	100.00	139	100.00	293	100.00

Source : Survey data

1. employment for longer days for all major technical combinations,
2. the increase in production of all crafts.

The re-organisation of fishing activity in the post-motorised period not only raised the production capacity but also resisted the deskilling of the work activity of the fishermen caused by capitalist intrusion. Further, while the new technology kept intact the traditional knowledge base, the re-organisation has given opportunities for re-skilling of work activity of fishermen.

Besides the production and reorganisation level, the changes in the capture fisheries have created impulses in the institutional setup like marketing; both product and credit markets.

The attempt at marginalisation of traditional fishermen by capitalist control of the capture fisheries were not confined to technology and organisation of production but were extended to create adverse market situations as well[2]. Capitalist sector while concentrating on high valued species 'get rid of the residue at throw away prices[3]. Further the bulk catch and the fresh supply by the capitalist sector caused the catch of the traditional sector to be of 'inferior varieties'. The domination of marketing set up by the capitalists made the middlemen to align with them thus completing the adverse locking system in which the traditional sector was cornered. These conditions have tremendously changed in the post-motorised scenario. The traditional sector with its increased production made its presence conspicuous in the arena of marketing.

By landing their catches in the conventional landing places of the motorised fishing crafts, they were transformed into marketing places thus destroying the capitalist marketing structure. Such scattering of the markets reduced the negative effects of dumping by the capitalist sector. The gradual spread of this marketing system enabled the fishermen to enjoy a sellers market. An increasing domination by the traditional sector has also brought some realignment of forces in favour of traditional fishermen. Middlemen's role in auctions which was shady during capitalist upper hand in production turned in favour of the fishermen. This change was visible at two levels. An exclusive set of middlemen emerged to deal with the motorised sector. Besides, they also began to offer increasing amounts as advances to the fishermen thus partially mitigating capital shortages faced by the traditional sector.

In the credit market, the massive investment requirement to undertake the transformation in the traditional fishery mainly came from private sector apart from contribution by co-operative ventures and government initiated schemes. The sample data give ample evidence to this. Table 7.4 shows the nature of financing of investment in fishing units in the motorised sector.

Table 7.4. Nature of Financing of Fishing Units

Sl. No.	*Nature of Finance*	*No. of Crafts*	*Percentage*	*Gears*	*Percentage*
1	Own Capital	10	3.41	27	9.22
2	Borrowings	42	14.33	80	27.30
3	Both	241	82.26	186	63.48
	Total	293	100.0	293	100.00

Source: Survey Data

Table 7.4 shows that more than 82 per cent of the crafts and 64 per cent of the gear were financed by borrowed and own funds. While a considerable segment of fishing units were financed exclusively by borrowings (14 per cent of craft and 27 per cent of gears) the share of own funds was negligible in the investment in the motorised sector.

To understand the credit market structure it is essential to know the sources of financing of investment Table 7.5 and 7.6 show the specifics of own and borrowed funds.

Table 7.5. Major Sources of Own Capital in the Financing of Investment in Crafts and Gear

Sl. No.	*Sources of Own Capital*	*Craft*		*Gear*	
		Number	*Percentage*	*Number*	*Percentage*
1	Ancestral gifts and savings	69	23.55	66	22.53
2	Savings	144	49.15	145	49.49
3	Savings and dowry	32	10.92	28	9.56
4	Ancestral gifts and dowry-	12	4.10	11	3.75
5	Saving and collective contribution	19	6.48	20	6.83
6	Ancestral gifts and profits from crafts	17	5.80	23	7.85
	Total	293	100.00	293	100.00

Source: Survey data

Table 7.6. Major Sources of Borrowings in the Financing of Investment in Crafts and Gear

SL No.	*Sources of Finance*	*Craft*		*Gear*	
		Number	*Percentage*	*Number*	*Percentage*
1.	Middlemen	107	36.5	141	48.00
2.	Bank	7	2.4	20	7.00
3.	Government Agencies	6	2.00	—	—
4.	Co-operatives	1	0.30	1	0.30
5.	Friends and relatives	—	—	14	4.8
6.	Middlemen and banks	113	38.6	98	33.4
7.	Middlemen and government agencies	57	19.50	17	5.80
8.	Banks and government agencies	2	0.70	2	0.70
	Total	293	100.00	293	100.00

Source: Survey data

On the basis of the nature of funds mobilised by fishermen for financing their capital requirements, it is seen that the major source of own funds by the fishermen was their savings. About 50 per cent of own funds emanated from savings followed by ancestral gifts and savings which was about 24 per cent of own, funds. It was also interesting that dowry and savings also constituted a considerable portion of self finance.

About the borrowed funds it was mainly the middlemen who formed the major financiers to the fishermen. Table 7.6 shows that 36.5 per cent of the borrowings were exclusively from the middlemen. More than 58 per cent of the crafts were financed by middlemen along with banks and government agencies. Even though banks and government agencies finance the crafts, such finances were insufficient to meet the purpose and hence the fishermen approach the middlemen to meet the deficit. In the case of gear, about 48 per cent of the finance was provided by the middlemen. Here the middlemen's role is dominant since another 39.2 per cent of the gear is financed by middlemen along with banks and government agencies. Thus in the post motorised fishing scenario, the augmented productive capacity assures flow of private capital to the fishermen in the absence of sufficient capital availability from other sources, particularly government agencies[4].

In the labour process analysis alienation of the workers from the work through extensive division and specialisation relating to technological advancement is a typical character of capitalist system. Apart from proletarianising, de-skilling due to fragmentation of work deprive the workers and their control over the work. But motorisation provided an opportunity for the empowerment of fishermen.

In the post-motorised setup, more and more fishermen became owners of productive equipments particularly through collective ownership. The sample data gives us information about the change in the ownership patterns. Table 7.7 portrays this.

Table 7.7. Economic Status of Fishermen During Pre and Post Motorized Period

SL No.	*Economic Status*	*Pre-motorsied period*		*Post-motorsied period*	
		Number	*Percentage*	*Number*	*Percentage*
1.	No Ownership	104	35.49	--	--
2.	Family Ownership	183	62.47	172	58.70
3.	Individual ownership	3	1.02	1	0.34
4.	Partnership	3	1.02	115	39.25
5.	Co-operative	-	-	5	1.71
	Total	293	100.00	293	100.00

Source: Survey Data

Data elicited from the fishermen reveal that 35 per cent of them were mere workers without any ownership right over any means of production. There were no opportunities for ownership in the pre-motorised era. This will be evident from table 7.7 where popular means of ownership such as co-operatives, partnership, individual ownership were either simply absent or constituted a negligible portion. The predominant type of ownership in the premotorised period was family ownership (62 per cent).

However, with motorisation new forms of ownership opportunities have emerged. Table 7.7 shows that those who were non - owners became owners of one type or other. Family ownership which is still the dominant type of ownership began waning from 62.47 to 59.70 per cent. Similarly, a notable change since motorisation was that the partner ownership of means of production grew up substantially (39.25 per cent). Co-operatives which was completely absent as a form of ownership in the pre-motorised sector made its presence however, marginally (1.71 per cent) under the new epoch. Individual ownership remained negligible in both periods.

Technology wise distribution of ownership is shown in table 7.8.

Table 7.8 enables us to understand the structure of ownership existing in the post motorised scenario in Kerala. It reveals that in the case of crafts requiring huge capital investment, the fishermen acquired ownership through partnership. Under this ownership pattern the fishing units are formed through the collective efforts ranging from 4 to 30 members. In the case of medium crafts where investments are less, it was the family ownership which helped fishermen to secure ownership rights. The non owners in the pre-motorised period became owners mainly through their collective efforts. More than $^{3}/4$th of the ring seine units were owned by fishermen through partnership. Thus the new technology enabled the fishermen community to enjoy ownership rights which give integrity to the work activity of fishermen.

The newer structure of ownership facilitated by motorisation engendered changes in the distribution of profits among fishermen. The workers share as a whole has increased. Table 7.9 shows the sharing system of profits existing under different technological alignments.

In the post- motorised period about $2/3^{rd}$ of the fishing units distributes 60-65 per cent of their profits among workers. Another 7 per cent distribute more than 70 per cent of their profit as wages among workers. The fact that partnership formed the main form of ownership of large crafts has resulted in majority of these crafts setting apart a major share ranging from 60-75 per cent as divisible income among worker. In the case of medium type fishing units like gillnet units 25 per cent of these crafts share 50-55 per cent of their profits as wages. Majority of these medium units (69 per cent) share 60-65 per cent of their profit as wages among the workers. Thus we see that in the post - motorised setup workers share as a whole has increased freeing them from capitalist exploitation[5]. Further the increasing accessibility to ownership helped them to regain control over the primary production.

In the post - motorised period, besides an increase in the sharing system of fishing workers, the employment potentiality of the sector has improved. In the mechanised sector, the average number of workers engaged in fishing per craft was 4, but in the motorised sector it was substantially higher. Table 7.10 shows the average worker per craft under different technical alignments.

Table 7.8 : Technology wise Distribution of Crafts in Terms of Ownership Pattern

Fishing days hours	*Dugout ring seine units*		*Plank ring units*		*Plywood ring seine units*		*Plank transom units*		*Gillnet units (large)*		*Gillnet units Small)*		*All units*	
	Number	*Percentage*	*Number*	*Percentage*	*Number*	*Percentage*	*Number*	*Percentage*	*Number*	*Percentage*	*Number*	*Percentage*	*Number*	*Percentage*
Family Ownership	1	5.88	16	28.07	1	10.00	31	100.00	31	79.49	92	66.19	172	58.70
Partnership	15	88.24	41	71.93	9	90.00	-	-	8	20.51	42	30.22	115	39.25
Individual Ownership	-	-	-	-	-	-	-	-	-	-	1	0.72	1	0.34
Cooperative Ownership	1	5.88	-	-	-	-	-	-	-	-	4	2.87	5	1.71
Total	17	100.00	57	100.00	10	100.00	31	100.00	39	100.00	139	100.00	293	100.00

Source : Survey data

Table 7.9 : Sharing Proportion Under Different Technical Alignments in the Post Motorised Phase.

Sharing Proportion	*Dugout ring seine units*		*Plank ring units*		*Plywood ring seine units*		*Plank transom units*		*Gillnet units (large)*		*Gillnet units Small)*		*All units*	
	Number	*Percentage*	*Number*	*Percentage*	*Number*	*Percentage*	*Number*	*Percentage*	*Number*	*Percentage*	*Number*	*Percentage*	*Number*	*Percentage*
< 50	-	-	-	-	-	-	-	-	-	-	-	-	-	-
50-55	-	-	-	-	-	-	25	80.65	18	46.15	27	19.42	70	23.89
60-65	10	58.82	56	98.25	7	70.00	6	19.35	20	51.28	103	74.10	202	68.94
70-75	6	35.30	1	1.75	3	30.00	-	-	1	2.57	9	6.48	20	6.83
>70	1	5.88	-	-	-	-	-	-	-	-	-	-	1	0.34
Total	17	100.00	57	100.00	10	100.00	31	100.00	39	100.00	139	100.00	293	100.00

Source : Survey Data

Table 7.10. Average Workers Employed Per Craft Under Different Technical Alignments.

Sl. No.	*Technical alignments*	*Number of workers per craft*
1.	Dug-out ring seine units	38
2.	Plank ring seine units	29
3.	Plywood ring seine units	37
4.	Plank transom units	3
5.	Gillnet units (L&M)	4
6.	Gillnet units (Small)	4

Source: Survey data

More work opportunities and increase in divisible income has improved the average income earned by fishermen[6]. Table 7.11 shows the average income earned by fishermen operating different fishing crafts. It reveals that considerable difference exists in average income earned by different types of crafts. A comparison of average income earned by fishermen as a whole in 1988-89 (Rs. 5,136) with that of 1996-97 (Rs. 24,400) shows that the income of fishermen has risen more than three and half times.

Table 7.11. Average Income Earned by Workers of Different Fishing Crafts.

Type of	*Dug-out ring seine units*	*Plank ring seine units*	*Plywood ring seine units*	*Plant transom units*	*Gillnet units (Large)*	*Gillnet units (Small)*	*All 'units*
Total income (in rupees)	2,37,93,900	3,67,63,440	58,54,140	18,23,580	63.81.040	90,99,864	8,37,15,964
fishing crafts							
Total workers	650	1,680	370	90	155	486	3,431
Average income	36,606	21,883	15,822	20,262	41,168	18,724	24,400

Source: Survey data

Thus we see that the new technology has facilitated a restructuring of capture fisheries both at productive and organisational level. These have improved the economic status of the fishermen in terms of income, employment accessibility to credit market and at ownership level. These are desirable developments; but it also gives reason for concern to the aritsanal sector. There are dark clouds over the horizon that can undermine the achievements in the artisanal fishery.

We have already established that the economic viability of the motorised sector was the making of a favourable price factor. The productive capacity of the traditional sector has increased substantially. In fact, there is an excess productive capacity. However, redundancy of capital emanates from a resource shortage rather than demand constraints. This proves the argument that Kerala fishery has reached the maximum of sustainable level of output. In this context, the augmented productive efficiency of the motorised sector indicates over investment in the sector. Resource shortage ends up with fishing attempts turning unprofitable and thus inflicting a heavy blow on the fragile economic build up of fishermen.

This situation has resulted in the emergence of certain capitalistic traits (apparently capitalistic features but camouflage the real problem) in the motorised sector. One such feature was fitting more OBMs in fishing crafts. In fact, fishermen are eager to raise the power of their units by fixing more engines on the belief that they could outrun their colleagues and catch more of precious little bounty. The study has looked into this aspect and table 7.12 shows trends in this regard.

In the motorised sector, more than ¾th of the crafts still operate with a single engine. However, there is a growing tendency among fishermen for more and more engines. While 14 per cent use two engines another 9 per cent use three engines or more. The temptation to use more engines are seen with crafts using ring seine. In the ring seine 47.62 per cent operate with two engines and 22.62 crafts operate with three engines. Among the crafts using ring seines, plank and plywood crafts are found to be using more engines. Another way the fishermen expect to catch more fish from the little available is using OBMs of high horse power, This trend is clear form table 7.13 where the horse power of the engines used by- different types of ring seine units are shown.

Table 7.12 : Distribution of Fishing Units According to Number of OBMS Possessed

Number of engines	*Dugout ring seine units*		*Plank ring units*		*Plywood ring seine units*		*Plank transom units*		*Gillnet units (large)*		*Gillnet units Small)*		*All units*	
	Number	*Percentage*	*Number*	*Percentage*	*Number*	*Percentage*	*Number*	*Percentage*	*Number*	*Percentage*	*Number*	*Percentage*	*Number*	*Percentage*
Single engine	9	52.94	6	10.53	-	-	31	100.00	39	100.00	139	100.00	224	76.45
Two engines	8	47.06	32	56.14	2	20.00	-	-	-	-	-	-	42	14.33
Three engines	-	-	15	26.32	3	30.00	-	-	-	-	-	-	18	6.15
More than three	-	-	4	7.02	5	50.00	-	-	-	-	-	9	3,07	
Total	17	100.00	57	100.00	10	100.00	31	100.00	39	100.00	139	100.00	293	100.00

Source : Survey data

Table 7.13. Distribution of Fishing Units According to Horse Power of Engines.

Horse power of engines	*Dug out ring seine units*	*Plank ring seine units*	*Ply wood ring seine units*	*Total*
15H.P	5	—	—	5
25H.P	15	75	12	102
40H.P	—	48	12	60
Total	20	123	24	167

Source: Survey data.

Table 7.13 shows that 61 per cent of the OBMs used by ring seine units are of 25 HP. and about 36 per cent are of 40 HP. It is sure that when existing OBM engines wear out, fishermen would go for high powered OBMs.

Excess capacity caused by resource scarcity is ominous; the run among fishermen for high powered engines and so on will swamp the traditional fishery with over investment. Such ventures would plunge the fishermen in to severe debt trap and to the ultimate disintegration and extinction of artisanal fishery.

The sample study has also focused on the major reasons of non fishing days. The study shows that non-fishing is forced upon by non availability of fish, even though some other factors too contribute to this. Table 7.14 shows the details in this regard.

Table 7.14 : Distribution of Fishing Units According to Major Reasons of Non Fishing Days.

SL No.	*Reasons*	*Number of units*	*Percentage*
1.	Bad weather	23	8
2.	Lack of Fish	240	82
3.	Equipment Repairs	15	5
4.	Lack of Crew	6	2
5.	Lack of Crew	9	3
	Total	293	100

Source: Survey data

It is evident from table 7.14 that scarcity of fish is the prime cause of non-fishing days and it confirms the magnitude of risks the fishermen are subjected to in the event of further over investment.

Motorisation has come up as a blessing to provide upliftment to the artisanal fisher folk. However, it is not an unmixed blessing- newer issues and problems are thrown up at them. This peculiar situation in artisanal fishing call for appropriate and ingenious measures. Development policies based on conventional paradigms have only aggravated the situation. We may have to design new strategies to solve the new challenges. Discussions with the experienced fishermen indicate the scope of new resource management techniques particularly with community participation to solve the fishery crisis.

In this regard, the study undertakes a critical assessment of the fishing policies in the next chapter.

NOTES

1. The SIFFS study of 1991 found that the average depth the motorized units fished during 1988-89 was 23 metres.
2. It may be recalled that under modernisation, importance was given to development of harbours and centres where mechanized boats could land their catch. This has obviously helped the capitalist fishermen.
3. It has been pointed out that the mechanized sector's sole concern was prawn catch. There was complete disregard for any other fish. A large number of juvenile fish, which could grow to a size of up to 25 kg are caught, though not wanted, to lie rotting or sold as "trash fish". This disregard by the capital sector amounts to destroyal of the mainstay of the artisanal sector (Iyengar, 1985).
4. Positive correlation between credit worthiness and flow of private capital has been pointed out by Platteau (Platteau, 1985). Even though Platteau had found this correlation in the mechanised sector, there is no reason to deny such a relation emerging in the motorised sector. The fact the money lenders are concerned only safe and productive lending of their capital. The emergence of middlemen in financing motorsied sector had thus assured sufficient flow of capital to the motorised sector.

5. In the mechanized sector, the share of the workers was steadily declined as a whole from 63 per cent to 43 per cent (see table 4.11) This showed that the increased exploitation of the workers under the mechanized sector.
6. In 1988-89, the SIFFS study has calculated the average income earned by fishermen working in the motorised sector in Kerala. The average income earned by fishermen of all categories calculated from this study was Rs. 5,136.

REFERENCES

1. Iyengar, V.L., (1985): Fisher People of Kerala, A Plea for Rational Growth", *Economic and Political Weekly,* (December), Bombay.
2. SIFFS (1990-91): *Motorisation of Fishing Units: Benefits and Burdens,* Summery Report of the Study, "Techno-Economic Analysis of Motorisation of Fishing Units Along the Lower South-West Coast of India, Trivandrum.

8

Major Fishery Issues and Shifting Policy Paradigms

Development of fishery has been an abiding concern of the government at the state and central levels. In the Kerala fisheries the role and perspective of Government in fishery development were ingrained in the laws formulated in this regard. Fishing is a traditional activity of coastal people since time immemorial. Since fishery operations were in its pristine forms and was undertaken by people of poor means as their livelihood, the customs evolved over the years could regulate the continuity of the operations in this sector obviating government regulations.

The British Parliament had passed Indian Fisheries Act - 1897 (Act IV of 1897) to regulate the fishing activity in British India. Later, the Cochin Fisheries Act 1092 M.E. (1917) and The Travancore Fisheries Act - 1097 M.E. (1922) were formulated specifically for development of fishery in Cochin and Travancore area. Similarly, in the Malabar Province Indian Fisheries (Madras Amendment) Act 1927 was also implemented. In 1949, after the unification of Travancore and Cochin to form the Travancore - Cochin State, the Travancore - Cochin Fisheries Act came into vogue.

Since independence, the realm of legislative approach to development issues became more concrete and participatory. Article 246 of the Constitution has provided the right to formulate laws in the territorial and inland waters and the right to initiate development of fishery to the state governments (Srivastava, et. al., 1991). This has given possibilities for more involvement in fishery development by state governments. The Union Government, in turn, was responsible for the development of fisheries beyond territorial waters and for maritime research works.

The laws in force prior to 1950 were oriented, inter alia, to achieve sustainable and gradual development of the fishery and protecting the interest of the real producers[1]. The basic premise of fisheries development plans in the Travancore region thus hinged on the judicious exploitation of marine resources by effectively and gradually raising the productive capabilities of the existing facilities and giving primacy to the accumulated skills of fishermen (Kurien, 1985). There were explicit regulations giving power to the state to control the size, type and number of crafts and gears and also prohibit fishing in part or full during peculiar situations. Stipulations were also there even to invoke licensing system in the matter of fishing process (Travancore - Cochin Fisheries Act -1950).

Development activities in the fisheries underwent a dramatic change since the inception of the Five Year Plans. Under the euphoria of planned development, all traditionally evolved concepts, formulations and equilibriums were replaced by western paradigms of development. The official view echoed in the National Planning Committee regarding fisheries sector was that it is an occupation "largely of a primitive character carried on by ignorant, unorganised and ill equipped fishermen. Their techniques are rudimentary, their tackle elementary, their capital equipment slight and inefficient" (Shah, 1948). This perception gave way to initiate modernisation of fishing sector on new lines. The method resorted to under this modernisation was the super-imposition of capitalist technology, foreign expertise and forging a link with foreign markets. Besides, this perception has led to the creation of a fishery bureaucracy and a string of scientific and research institutions to form the basis for modernization of the sector. These strategies expected to improve the socio-economic conditions of the artisanal fishermen. The inconsistency and contradictions of this policy which culminated in uprooting the artisanal fishermen is discussed elsewhere. The capitalist forces unleashed under the western concepts of modernisation undermined the otherwise strategic path of development such as co-operatives[2]. Even though, there was a good legacy of co-operative efforts to ameliorate the economic lot of the fishermen, this institution was the first casualty of the modernized approach of development in fishery[3]. The confidence of the authorities in the modernisation process was such that the market principles would be taken care of the whole issues, and they just ignored the importance of co-operative endeavours in the fishermen development. However, the inconsistencies and contradictions caused by the modernisation process in the artisanal sector necessitated some policy changes and subsequently some patch up measures were introduced. There were no scope for much flexibility of policies under the given situation and hence looked upon the co-operative approaches once again[4]. Thus the co-operatives as a strategy of development in fishery which was ignored completely under the First Five Year Plan was given importance in the Second Five Year Plan. In 1958-59 the State Government initiated steps for organisation of a three tier system of fishery co-operatives. Still the assumption was that once the co-operatives were established, productive equipments could be given to them to augment their productive capacity which in turn will result in creation of incessant surplus capable of transforming the entire artisanal fishery into modern (capitalist) sector.

It is obvious that such adhoc measures could not succeed[5]. At the policy level this adhocism failed because of:

1. only meagre provisions were given in organising co-operative units. Out of the total State Plan funds of Rs.31.51 crores invested in the fisheries sector upto 1979 - 80 the expenditure incurred on fishery co-operatives were only Rs.2.39 crores (7.6 per cent). The planwise break up is shown in table 8.1.

Table 8.1: Planwise Breakup of Expenditure Incurred on Fisheries Co-operatives. (Rs. lakhs)

Plan period	*Total state expenditure on fisheries*	*Expenditure on fishermen co-operatives*	*Percentage of expenditure on co-operatives of the total*
1951-56	2.74	-	-
1956-61	60.52	6.85	11.3
1961-66	343.24	11.05	3.2
1966-69	749.33	25.38	3.4
1969-74	563.38	54.34	9.6
1974-78	782.96	85.75	11.0
1978-79	279.69	37.62	13.5
1979-80	369.21	17.56	4.8
Total	3151.01	238.55	7.6

Source: Department of fisheries, (1986)

2. the institutional set up in the modern sector which was capitalistic in nature (credit, marketing, productive efficiency) put the traditional fishermen at a disadvantageous position vis-à-vis the modern sector and at this competitive level, the co-operative endeavours in the fishing sector could not be a viable proposition. The development efforts under the Five Year Plans allocated more funds to help the capitalist forces[6].
3. further, the increasing role of merchants/middlemen under the capitalist development process resulted in confiscating all benefits of the real fishermen.

Another policy which the Government favoured was an export boost of the fishery products. Under this policy fishery development was equated with development of trawling. Government loans and subsidies and bank credit were channelled to raise export markets of fish products. Almost the entire plan investment was for development of trawling and infrastructure support for it. Here again, the underlying expectation was that export earnings would facilitate the socio-economic upliftment of the fishermen.

The burden of the misdirected policies which culminated in the severe deprivation of the artisanal fishermen induced them to collectively air their grievences. The artisanal fishermen, whose knowledge of fisheries extends to centuries of experiences and observations through generations, diagonised that the basic maladies afflicting in the fishery sector was the outcome of lopsided development approaches and strategies pursued under modernisation attempt. They collectively articulated the problems of fishery and demanded for :

1. conservation of living resources in the coastal sea,
2. regulation of indiscriminate fishing by mechanised boats, and
3. protection of artisanal units from the onslaught of the modern sector.

Also, direct action of fishermen described above was to ensure their rightful place in the formulation and implementation of fishery development. By and large, the demands of artisanal fishermen reflected the crisis in the fishery sector. It just not ceased as a problem of fishermen alone. The lopsided development of the fishery and the isolation of fishermen from the development efforts and, moreover, the inseparable link between the fishermen and the sea invited the attention of social activists, thinkers and voluntary organisations (Kurien, 1988). All these agencies showed their solidarity for the cause of fishermen through extensive studies, seminars and writings. These had resulted in identifying various issues as pertinent problems facing fishery which require immediate intervention by the Government. The major fishery issues which required immediate intervention by the Government were:

1. over exploitation of coastal fisheries necessitating imminent conservation and management of fishery resources.
2. protection of artisanal sector, ensuring a fair share of the resources.
3. optimisation of the size and power of the fishing units in the artisanal sector for operation within the inshore sea to avoid excess capacity and over investment in the artisanal sector.
4. periodical assessment of resources and drafting of appropriate fishing programme to suit the resource capabilities.
5. enactment of laws to provide the right of first sale to the primary producers to save them from exploitation by auctioneers and middlemen.
6. inducement to fishermen for offshore fishing with viable technology.
7. provision of liberalised and adequate organised credit.
8. provision of adequate supply of spares of engine and service facilities.
9. identification of forward and backward linkages in the fishing process to augment employment opportunities to the fisher folk.
10. indiscriminate construction of fishing harbours to meet the demand of mechanised boats at unrealistic projections.
11. over exploitation of deep sea resources by multinational companies endangering the resources in the inshore sea affecting artisanal fishermen.

The mounting pressure of the fishermen and the society at large up on the Government on identification of specific issues necessitated the Government to be empathetic with the fishermen cause. The Government realised the pitfalls of the policies pursued in fishery and began to contemplate on new strategies and policies that could accommodate the interest of the real producers. The Government started a multi-pronged strategy to deal with the situation.

The Government swung into action by enacting certain legislative measures. The Kerala Marine Fishing Regulation Act (1980) provided for regulation of fishing by the mechanised boats, registration and licensing of all boats and demarcation of the coastal waters (upto 30 metres south of Quilon and 20 metres depth north of it) for the exclusive use of artisanal fishing craft. The underlying objective of the law was conservation of marine fishery resources.

Another piece of legislation which the Government brought forth was The Kerala Fishermen Welfare Funds Act (1985). This is intended to usher in new vistas of funding support to fishermen's welfare measures such as health cover, marriage and death ceremonies, old age care, short term credit for consumer expenditure and education. In the meantime, Government also appointed some committees to augment its own information base regarding fisheries[7]. The Kalawar Committee (1984) whose assigned task was to examine the impact of trawling during the monsoon season on shrimp resources and particularly its impact on the traditional sector, among other things, pointed out the necessity of restricting the operations of the mechanised sector to augment the productive capacity of the artisanal fishery.

Armed with, the legislative powers and the information base, the Government initiated a number of conservation, regulatory and welfare measures. The government pursued rigorously certain programmes favourable to artisanal sector. The Government's earnestness to favour artisanal sector was reflected in the Seventh Five Year Plan outlay set apart for artisanal fishery development. Out of the total outlay of Rs.40 crores, about Rs.15 crores was set apart exclusively for artisanal fishing as against Rs. 0.60 crore in the Sixth Plan. A major portion of this outlay was used to supply motorised crafts, FRP boats, beach landing crafts and selected gear. An elaborate programme was also visualised for improving the infrastructure facilities for primary marketing of fish to ensure better prices for their catch and liberating them from the claws of the middlemen.

Further, the Government was saved of the trouble by finding alternative strategies to suit the artisanal sector. The fishermen have already articulated their preference for conservation and augmentation of resource base and raising harvesting capacity through motorising country crafts[8]. The Government simply need to clinch this opportunity centralising all their efforts in this line. It was also increasingly felt that there could be a nodal agency to co-ordinate the development activities in the artisanal sector. Bitter experience with the co-operative movements to improve the economic condition of the artisanal fishermen made to accept the recommendations of the Resuscitative Committee on Fishery Co-operatives[9] to formulate Fishermen Welfare Societies. The society will function as the central agency for supplying all inputs for fishermen, implement village level programmes for the promotion of all round socio-economic upliftment of the community. Besides, it will be the grass root level agency for planning and plan implementation in the fishery sector.

At the district level, District Co-operative Societies and at state level Kerala State Co-operative Federation for Fisheries Development as an apex body have been formulated for better co-ordination and functioning of the primary organisations.

A prime conservation measure adopted by the government was the ban of trawling. Even though the government had the recommendations of two earlier committees, it constituted a third expert committee to re-examine the question of ban on monsoon trawling[10]. This committee submitted its report to the government in June 1989 which had recommended imposing trawling ban for three months. Following this recommendation, the government has imposed for the first time a total monsoon trawling ban from mid July to the end of August[11]. Thereafter, the trawling ban became a permanent fishery management practice in Kerala fishery [12]. This will be evident from table 8.1.

Table 8.2. Number of Days of Trawling Ban During 1988-96

Years	*Period*	*Number of days*
1988	June 29 -August 31	64
1989	July 20 - August 3 1	43
1990	June 28 -July 21	24
1991	July 15 - August 16	33
1992	June 21 -August 3	44
1993	June 15 -July 15	31
1994	June 15 -July 29	4
1995	June 10 -July 20	41
1996	June 15 -July 25	41
1997	June 15 -July 29	45
1998	June 15 -July 29	45

Source: *Achari, et. al, (1995), Stanley, (1996) and Deepika* Daily (1997 & 1998)

The effect of trawling ban on conservation of resources is an unsettled issue. However, in 1990, the government constituted an inter-disciplinary study team to assess the impact of monsoon trawling ban. The team found that the fish harvest in 1989 was a record level of 6,40,000 tons, 170,000 tons above the 1988 level. In 1988 itself when there was partial ban, the fish harvest for the year was up by over 80,000 tons of the 1987 level. Considering the fact that fish resources are influenced by a multiplicity of factors, it may be difficult to quantify the effect of ban on conservation of these resources. However, having well settled the fact that trawling process amounts to ecological destruction, a ban would do good for fishery resources and in particular the artisanal fishery[13].

The declaration of the fisheries policy in 1994[14] was a mile stone in the fishery development of the state. The policy of 1994 is significant that the government has understood the real causes of the fishery crisis in Kerala and recognizes the need for certain fundamental principles of growth that should be pursued in a Third World economy like India. The policy statement in its preface remarked that the present state of affairs in the fisheries was due to the lack of well defined declared state policy. Further, it has pointed out that a socio-economic dichotomy has emerged in the sector.

The basic premises of the Fisheries Development and Management Policy Statement emerged from the realization that the modernisation process rooted in high technology, capital intensity and foreign markets could not succeed in giving an improved standard of living to the fishermen. The policy further reckons that modernisation only divided the sector into the polarised classes of haves and have nots. The inappropriate modernisation process has caused, apart from this socio-economic changes, a fall in productive efficiency on account of a stagnant production on the one hand and increased in production expenses on the other. These effects are pointers to an irrational use of scarce resources of the marine economy. It also undermined the end purpose of the functioning of this sector by depriving low cost protein to the society at large (due to excess importance given to export). It is to be noted with relief that the policy gives due importance to traditionally evolved knowledge base and remorsefully accepts that the indifference shown to traditional paradigms had aided and abetted polarisation of fishery sector.

Based on these, the New Fishery Policy envisaged a set of goals to recover the lost dynamics of the fishery sector. The important goals set out in the policy are (Govt. of Kerala, 1993:).

1. sustainable development of fishery resources.
2. improvement of standard of living of fish workers.
3. ensure availability of fish resources both for domestic consumption and for exports.
4. continuing welfare activities for fishermen development.

The major programmes that are to be implemented to secure the objectives stated in the policy statement are:

1. to accord fish production and fish processing the status of agriculture and industry respectively. This will entitle the fishery sector to be eligible to receive all assistance/subsidies recommended for agriculture and industry from time to time.
2. to focus on artisanal fishermen who are engaged in fishing and fish related activities as special target group while implementing the fishery policy.
3. to make periodic assessment of the fishery resources available to the state so as to evolve sustainable fishing efforts.
4. to formulate an 'Aquarian Reform' relating to coastal waters of the state with an objective of ensuring ownership rights of fishing artifacts exclusively to real fishermen.
5. to encourage development of technology appropriate to the socio-economic conditions of the fishermen and to the state.
6. to limit the investment sufficient to exploit the fishery resources at a sustainable level. The technique of participatory management is to be used (fishermen's participation) in such endeavours.
7. formation of an export policy emphasizing value added export without depriving fish availability to the domestic consumers.
8. ensure through appropriate legislation the right of first sale to the primary producers to save them from middlemen.
9. formation of a development and an infrastructure policy keeping into account the decentralised pattern of growth.

The Fishery policy is progressive as it strives to promote integrated development of the fishery. Further, the policy give recognition to the ideas of development which the real producers had in their mind and for which they had fought for more than two decades. In other words the government through this policy has concurred with the contention that the development paradigms must evolve organically from the internal dynamics of the sector rather than simply copying western paradigms. In fact, this was the pattern of development that took place in developed economies.

In sum, the labour process changes in LDCs give natural expression to certain unique development formulations which hinge upon the forces and stimulants of development that spin off from the internal dynamics. Any development or policy paradigms inconsistent with such organic forces would cause social tensions. Kerala fishery is a typical example. The experience in Kerala fishery shows that such conflicts and tensions serve as a corrective force to bring in a more realistic development approach that would make development, participatory, equitable and sustainable. The following conclusions emerge from this analytical study of technology and labour process changes in the Kerala fishery.

1. Capitalist forces which are ingrained in modern development paradigms do not always work to help the LDCs as visualised by the modernists.
2. Development forces in a sector/society/economy must emerge from its internal dynamics.
3. Development agencies such as government should permit such forces of development to crystalise into institutions and arrangements and provide facilities for evolution and growth of such forces.
4. Ignoring information and knowledge base acquired by people over the years would be to the peril of the real development forces.

NOTES

1. The tone of the laws were generally affirmed the importance of nurture fishery strategies. It recognised the time needed for stocks to replenish themselves, the need to conserve species diversity, the use of a range of selective techniques to take a seasonally diverse catch.
2. The importance of co-operative efforts to improve the economic conditions of fishermen was realised at the beginning of this century. In Kerala, the first fishermen society came into existence in 1917. By 1933, there were about 95 societies which were

primarily functioned as credit societies. To strengthen such co-operatives the Govt. of Travancore in 1934 was advised by a Committee (Paramupillai, 1935) to convert them as multipurpose co-operatives and providing provisions of processing facilities such as curing yards and involvement of community leaders and constant Government support. The concept of 'co-operatives' thus involved a well thought out and integrated set of policies keeping the real producers at the central focus.

3. Apart from confidence in other capitalist principles (disguising them as same as growth principles/strategies) the modernisation attempts in the fishery impliedly believed in Lewisian type of development approach (Lewis, 1955). The whole concept of the development under the modernisation process was that the new height of capital regime would generate enough surplus by the fishermen which would be ploughed back and the process to continue until all fishermen were brought in the ambit of modernisation!
4. A general feature of these co-operatives was that they created from above and handed down to fishers, quite irrational to the spirit of co-operativism.
5. In a study about the co-operatives and mechanisation and their impact on traditional fishermen, it was found that the benefits of mechanisation was usurped by a group of people who had set up fictitious fishery co-operatives (Hakim, 1980).
6. In a study on credit and indebtedness among the marine fishermen of Southern Kerala, Plateau et. al, showed that the volume of credit and other institutional borrowings were positively correlated with the degree of mechanisation. Further, it was found that while major part of borrowings in the mechanised area was for investment, it was for consumption in the artisanal sector (Platteau. et. al, 1985).
7. Important Committees appointed by Government in fisheries were
    1. Babu Paul Committee (1981)
    2. Kalawar Committee (1984)
8. Attempts to motorise the country crafts were done by fishermen individually and collectively at different hamlets of the coastal

area. Some fishermen groups were succeeded in converting the country crafts to fix out board engines. This invention/adaptation wide spread all over the coastal area with in short time;. Both Central and State Government policies also helped substantially to intensify this trend.

9. Resucitative Committee on Fishing Co-operatives was constituted in 1975 to enquire into the failure of co-operative movement in fisheries.
10. Balakrishnan Nair Committee (1989)
11. In 1988, following an agitation threat by fish worker's unions (mainly Kerala Swathanthra Matsya Thozhilali Federation) the government had promulgated a partial ban by which all the trawler operating centres in the state except Neendakara, the largest centre in the state were ordered closed for the months of July and August.
12. Clamping of monsoon trawler bans was not an easy task for the State government since it involved a decision against the strong capitalist interest in the fishery and also at times against the Central Government's policies. The economics and politics of the trawler bans was discussed by John Kurien (Kurien, 1991).
13. Scientific opinion had confirmed that trawling is an ecologically destructive process. After 1952, when trawling had been introduced in the Indo -Norway region Project, it was banned in Norway. In 1976, trawling was banned in the Philippines since it destroys juvenile fish as well as organisms which fish feed on. In 1979, the University of Philippines and the College of Fisheries studied the impact of the ban and found that "In one year the biomass over all depths had doubled, while the most seriously affected depth range of 10 to 58m recovered to the extent that the biomass increased by more than 100 per cent". The study stated that the imposition of a trawling ban is a suitable tool in tropical waters to protect heavily exploited fish stocks to recover. The destructive process has been observed in Kerala also. According to experts, the life span of Prawn Stylifera is two years and it grows upto 110mm. Analysis of catch during monsoon seasons shows prawn sizes of 51-98mm. Besides many varieties of prawn like the highly prized prawn Indicus became extinct (Iyengar, 1985).

14. The State Government constituted a high powered committee in October 1992 comprising Fisheries Secretary as its Chairman and Fisheries Director as Convener. The Committee also included fishery scientists and experts in the socio-economic field. The committee submitted a draft report in April 1993.

REFERENCES

1. Achari, et al., (1995): *Fisheries Development in India, Problems and Prospects,* PCO, Trivandrum.
2. Govt. of Kerala, (1993): *Fisheries Development and Management Policy.*
3. Govt. of Kerala, (1976): *Report of the Resuscitative Committee for Fisheries Co-operatives,* Government of Kerala, Thiruvananthapuram.
4. Hakkim, A.V.M., (1980): *Mechanisation and Co-operative Organisation. Their Impact on Traditional Fishermen of Kerala.* (Unpublished M.Phil discertation. CDS Trivandrum.
5. Iyengar, L. Viswapriya, (1985): "Fisher people of Kerala, A Plea for Rational Growth", *Economic and Political Weekly,* (December), Bombay.
6. Kalawar, A.G. et al., (1985): *Report of Expert Committee on Marine Fisheries in Kerala, 1982,* Central Institute of Fisheries Education, Bombay.
7. Kurien, John, (1985): Fisheries Development, The Experience of Norway and her Partner Countries within the Context of Norwegian Assistance", Proceedings of the *International Seminar of Fisheries Development,* Ian Bryceson (Ed), Royal Norwegian Ministry of Development Corporation, Oslo.
8. Kurien, John, (1988): "The Role of Fishermen's Organisations in Fisheries Management of Developing Countries", in *Studies On The Role of Fishermen's Organisations In Fisheries Management.* Fishing Technical Paper 300, FAO, Rome.
9. Kurien, John, (1991): "Ruining the Commons and Responses of the Commoners: Coastal Over fishing and Fishermen's Actions in Kerala State, India", *Discussion Paper 23,* UN Research Institute for Social Development, Switzerland.
10. Lewis, W., Arthur, (1955): *The Theory of Economic Growth,* Alien and Unwin, Ltd. London.

11. Paramupillai, K., (1934): *Travancore Co-operative Enquiry Committee Report,* Thiruvananthapuram, (in Malayalam.

12. Paul, Babu, (1984): *Report of the Committee to Study the Need for Conservation of Marine Fishery Resources During Certain Seasons of the Year and* Allied Matters, Government of Kerala.

13. Platteau, J.P., et al., (1985): *Technology, Credit and Indebtedness in Marine Fishing, A Case Study of Three Villages in South Kerala,* Hindustan Publishing Corporation, Delhi.

14. Shah, K.T., (1948): *National Planning Committee : Animal Husbandry, Dairying , Fisheries and Horticulture,* Vora and Co., Bombay, 1948.

15. Srivastava, U.K., et al., (1991):*Fishery Sector of India*, Oxford and IRH Publishing Co. Pvt. Ltd.,. New Delhi.

16. Stanley, S., (1996): *Impact of Trawing Ban on Primary Sector with Special Reference to Sakthikulangara Panchayat* (Unpublished M.Phil Dissertation), University of Kerala, Thiruvananthapuram.

9
Summary and Conclusions

Fisheries sector in Kerala has experienced certain unusual changes. The modernisation of fishery sector in Kerala was ushered in with foreign assistance in 1953. The primary objective of this modernisation attempt was raising the standard of living of the fishermen community by augmenting their productivity. Earning more foreign exchange and raising the domestic availability of fish were also envisaged.

Three decades of development experience in the fishery portray that the economic condition of the real fishermen deteriorated and their standard of living declined. The per capita income of the fishermen community stagnated below the state average. Their means of production became insignificant and irrelevant at operational level and work opportunities of the fishermen declined. In short, they were almost thrown out from fishing activity which had provided a source of livelihood for a long time.

In fact, at a broader level, the development experience of Indian economy for more than half a century also showed similar experiences. Like in the fishery, India's planned development too could not attain its stated objectives. More than that, the development course had manifested forces that aggravated the basic maladies in the economy. These facts highlight that at micro and macro levels, development issues have a common pattern. It indicate that something is at fault with the development approach we pursue.

The successful development experience of advanced capitalist countries had provided substance for the formulation of compendious models of growth. Their experience taught us to believe that such crisp profiles would be a panacea for all the economic ills. India's long colonial association with such economies has facilitated a quick and easy transplantation of these paradigms in LDCs. Limited education on western lines and imitation of similar values and attitudes and the corresponding hate of the intelligentsia towards indigenous systems culminated in to an outlook that there is no alternative to the western paradigms of development.

This flawed perception made us to lose sight of an original, organic and independent growth pattern conforming to the socio-economic realities in LDCs. Major theories which criticised the capitalist pattern of growth too could not offer a growth perspective which could draw its propulsion from internal dynamics of the economy. Instead, such theories preferred to see development problems of LDCs as problems of social relations which could be settled at the best through a class war. Some economists who were also drawn to their tools of analysis from class clevages, however, gave a more realistic perception to the development issues of LDCs. Their vision that the present predicament of LDCs are the inevitable outcome of capitalist growth have done a great deal in bringing development issues of LDCs as a distinct problem. This qualitative shift in attitude towards the issues of LDCs had prompted many writers to focus critically on the relation between developed economies and less developed ones. Many have found that the LDCs are beset with a series of structural contradictions emanating from the dependant relationship particularly from technological dependency.

Technological dependency compels us to ignore the socio-economic context of its formation and use. We ignore the fact that the leaps in technology is driven out by profit motive and not by any societial considerations. LDCs fervently welcome technology transfers but remain oblivious of the socio-economic disequilibrium it sets in. While the profit designs of the capitalists at the centre keep up upshots in technology, the profit clamour of the capitalists in the periphery provide the way for entry of modern technologies in LDCs. Thus a coalescence of the metropolitan and the periphery capitalists and the perspective that there is no alternative to western paradigms of growth provide an avenue for the entry of foreign technology in LDCs causing vicious circles of technological dependency on the one hand and consequent dependent relations in many areas on the other.

The view that technology is always progressive deny the opportunity to critically assess the implications of its use. We should learn to distinguish profit considerations of alien technology from that of its socio-economic implications. Writers like Braverman have pointed out the necessity of looking technology from a societal perspective rather than from its engineering dimension. The dependent relation of LDCs, and its structural disequilibriums necessitate the assessment of technology from societal premises. The concept of labour process constitute a typical tool to focus on the simultaneous interaction between the technology and the society. The labour process analysis encompasses technological changes and its implication upon the means of production, its organisation and the subsequent class polarisations. It provides a neat canvas to comprehend all such changes in LDCs. In the context of dependent relations its use provide scope for understanding newer upshots in the societial changes. It is these dimensions that we have used to explain the unusual changes that have occurred in fishery.

In LDCs the labour process analysis amounts to a conflicting relation between two modes of production because of its dependent relation. Unlike in capitalist economies where the articulation between modes of production progressed into a capitalist phase, such a linear change could not occur in a less developed economy. This is because a complete transformation to capitalist epoch could not occur in LDCs on account of limitation of resources. The resource transfer which occurred in the previous centuries deprived an original and organic growth in LDCs. Its continuation in the present global regime robe the resources required for a full capitalist growth. The partial transformation, in fact, is an outcome the capitalist countries created in LDCs through the western paradigms of development. Such situations will be maintained because metropolitan capitalist did not favour a change. Also, resources are still drawn away making it impossible for a neat and complete transformation.

In the fishery development it was held that Kerala fishery is logged in slumber mainly due to inadequate technology. Capitalist development outlook viewed that the production process that evolved over a long time in the fishery is insignificant and irrelevant. Through INP in 1953, the foreign agencies who were the propagators of this new development design found a foot hold in Kerala fishery. They found that the means of production which the fishermen community were accustomed to since a long time were incapable of modernisation. Consequently some foreign models were brought in and replicated in Kerala fishery. In the meantime, the discovery of foreign markets for some marine food products had resulted in specialising in such products and leaving the general fishery development in the lurch. The foreign intervention, the subsequent dissemination of new technology and the opening up of lucrative foreign markets brought the capitalist forces to the threshold of unlimited expansion of the fishery.

An embodied form of technology and an official rhetoric on modernisation have made capitalist entry an easy affair. The study has identified some specific ways through which the capitalists established their sway in the fishery:

1. they have created a condition for emergence of wage labour.
2. they increasingly began to control work activity through de-skilling traditional knowledge system.
3. they clinched the institutional credit and government policies in their favour,
4. they achieved domination in markets.
5. exploited workers to squeeze out profits by keeping the crew share unchanged for more than three decades.

The capitalist production process had accentuated the down fall of the artisanal sector.

The modem sector rendered the traditional technology less and less viable which has manifested in declining share by the artisanal sector (Production of the traditional sector continuously declined from 100 per cent in 1956-59 to 97 percent in 1960-66, to 84 per cent in 1967-75, to 69 per cent in 1976-80). Similarly, environmental factors had also caused a considerable fall in the productive capacity of the traditional sector. All these resulted in permanently blocking the artisanal sector growth. This situation in the fishery has proved our theoretical contention that its peculiar socio-economic condition did not permit LDCs to transform its attempt of development into a fully developed capitalist growth. The subsequent changes in the fishery, further confirm our notion that given such incomplete transformations, there will be, attempts of inevitable changes emanating from the internal dynamics of the economy.

This study has found that the deprivation of the fishermen infused them to shed off their initial passivity and swing into action, actively and vigorously. This reaction resulted in two dimensions: while the former was the usual resort to collective bargaining, the latter was a unique collective effort validating our theoretical upshots. In collective actions of the fisher folk, we found certain specialties. Their initial group activities were either localized or restricted within their religious dogmas. However, the capitalist pressure intensified their group activities to surpass all religious and local inhibitions. We also have observed a total change in the character and motives of collective efforts of the fishermen. While the collective efforts have provided an opportunity of scientific orientation and articulation of their perspective, changes they brought at the production level was profound. The fisher folk started intervening with technical innovations and adaptations. The fishermen attempted their technological intervention at:

(a) development of knowledge of fixing OBMs in the country crafts,
(b) making of plywood boats,
(c) fabrication of more efficient gears,
(d) changes in organisation of production, and,
(e) construction of artificial reefs and use of fish attracting lanterns to augment fish production.

The conflicting terrain which has developed in the fishery between the capitalist modernists and aritsanal fishermen has produced amazing changes in the Kerala fishery since 1980s. This has been treated as a separate epoch known as 'motorisation' as against 'mechanisation' of the earlier three decades. This motorisation which was an outcome of labour class response against capitalist intrusion has set in newer changes. A cardinal change was that the artisanal fishermen succeeded in retrieving dominance in their traditional bastion.

An attempt for an assessment of the motorisation is made with the use of primary data. It reveals that the fisher folk staged a come back mainly by augmenting their productive capacity. The new technology helped them to overcome many of their infirmities suffered under capitalist mechanisation. The data showed that their productive capacity have been enhanced through:

1. lengthening the sea faring capacity,
2 raising their fishing time, and,
3. increasing the number of fishing days.

It has been found that about 56 per cent of the crafts are now able to fish 35 meters and more as against an average 11.8 meters in the non - motorised sector. Similarly, 71 per cent of crafts fish between 6 - 8 hours against a mean time of 3.28 in the non mechanised category. Also, 63 per cent of crafts now fish more than 200 days which is far ahead of the fishing days of the non motorised sector.

The heightened productive capacity also became instrumental in attracting abundant capital into the artisanal fishery. This is borne out from the data that 82 per cent of crafts and 64 per cent of accessories including gears were financed mainly by borrowings supplemented with own savings.

The motorisation has brought some favorable changes in the ownership pattern as 40 per cent of non-owners became owners during post motorised period mainly through partnership and co-operative ownership. This increasing ownership rights and control of fishery by artisanal fishermen have resulted in raising the divisible income in favour of fishermen. It was found that 2/3 of the fishing units distributes 60 - 65 per cent of net receipts as wages. In short, all these changes helped the artisanal fishermen in recapturing the labour process.

An analysis of the motorisation process in terms of economic performance reveals that, by and large, this epoch has been a success as majority of these crafts are making profits. While in the North zone it was 87 per cent, in the South, 91 per cent of the crafts are profitable.

In terms of rate of return on capital, a comparative analysis shows that the Southern fishery gets a return of 63 per cent as against 34 per cent in the North. It was found that this difference stems mainly on account of higher operating expenditure, sales commission and crew remuneration in the Northern fishery.

A measurement of physical efficiency of production in terms of catch per unit effort and catch per unit energy shows that in the Kerala fishery as a whole, these indices are 2.85 kg and 0.26 kg respectively. Zone wise comparison of this shows that both are almost identical.

Post motorised phase has put out certain interesting conclusions. While motorisation is an attempt by fisher folk against capitalist exploitation, it has not augmented the productive efficiency of the fishery as is evident from the low values of catch per unit effort and energy, over time, in comparison. However, fishing remain a profitable venture on account of a favourable price factor, which remain a notable feature of Kerala fishery.

Similarly, it has been observed that the artisanal fishermen who were critical of capitalists production themselves are increasingly pursuing capitalist behaviour. More crafts fixed with high powered OBMs are introduced, the size of the nets has increased, mesh size narrowed and the number of OBMs in crafts have raised. However, the pursuance of the capitalist production traits could not be construed as offshoot of animal spirits of competition. On the other hand this points to the fact that Kerala fishery is still affected by the remnants of the ravages of capitalist growth. Marine production has reached the pinnacle heights of maximum sustainable yield under the mechanisation phase itself. Motorisation process has inflated the number of crafts bringing down catch per unit effort and compelling the fishermen to put in more fishing efforts and thus making fishing activity more and more expensive. This results in over investment in fishing units and at the point where catches could not pay for such investment, a vicious circle sets in making fishing catastrophic to artisanal fishermen. Given this impending scenario, we endeavoured at an assessment of government policies to highlight the policy changes required to maintain the advantage of the artisanal fishermen in their protracted struggle against the capitalists.

A realistic approach to fishery development was conceived by government of Travancore as part of modernisation during pre-independence period. The prime thrust in this attempt was to augment the productive capacity of the artisanal sector through gradual improvement of the means of production. However, this gradual approach was viewed as 'backward' under the premises of capitalist development. At organizational level, co-operatives were also formed for all round benefit of fishermen. The policies since independence actually resulted in thwarting a linear and organic growth process embedded in earlier development approach. The principle of co-operatives was tried to help the fisherman to withstand the capitalist pressure. Capitalist succeeded in scuttling such attempts of empowerment and hijacked the systems in their favour. The plan allocations also aided the strengthening of the capitalist forces in the sector.

The fishery crises since mid 1970s reflected the uncertainty and lack of proper direction in government policies. The scientific temper and tone of the articulation of fishing issues by the artisanal fishermen as well as by expert committees, however, has provided some sense of policy direction to the government. Government was convinced that the crisis in the fishery was attributed to over fishing resorted to by mechanized and motorised sectors and thus driving home the need for conservation of resources and initiating sustainable development paradigms. This is increasingly reflected in various provisions and programmes envisaged in the Fisheries Policy formulated by the Government since 1994.

Thus on balance, in the Kerala fishery we see a shadow of complex issues still existing as remnants of capitalist growth. The survival strategies which have emerged as a response against isolation and deprivation of fishermen community, per se could not save the fishery. However, such strategies indicate that future issues are reflecting to newer resource management techniques involving more and more of community participation to secure sustainable development of the fishery resources. At a general level, the development experience of the fishery teach that the process of modernisation should not be imposed vertically; it must be propelled from its internal dynamics. A lesson of profound importance is that traditional knowledge base and experience nurtured and enriched by generations should not be dubbed as irrelevant and inappropriate; rather strategies of modernisation must draw up such knowledge systems as infrastructural columns.

Bibliography

1. Adelman, I., and C.T. Morris, (1973) : *Economic Growth and Social Equity in Developing Countries,* Stanford University Press.
2. Amin, Samir, (1974): *Accumulation on a World Scale Vol.I New York.*
3. Ashworth, William, (1987): *A Short History of the International Economy Since 1850,* Longman Group UK Limited, London.
4. Bagchi, A.K., (1972): "Some International Foundation of Capitalist Growth and Underdevelopment", *Economic and Political Weekly,* Special Number, November.
5. Bagchi, Kumar, (1995): *New Technology and the Worker's Response, Microelectronics, Labour and Society,* Sage Publications, New Delhi.
6. Bain, J.S., (1956): *Barriers to New Competition,* Harvard University Press.
7. Baran, Paul, (1962): *The Political Economy of Growth,* New Delhi.
8. Bhalla, A.S., (1965): "Choosing Techniques: Hand Pounding, V. Machine Milling of Rice: An Indian Case", *Oxford Economic Papers,* Vol. 17
9. Bhalla, A.S., F. Stewart, (1976): "International Action for Appropriate and Technology", *Background Paper for the World Employment Conference,* Geneva: I.L.O.
10. Bharadwaj, Krishna, (1994): *Accumulation, Exchange and Development, Essay on Indian Economy,* Sage Publications, New Delhi.
11. Boserup, Ester, (1981): *Population and Technology* Basil Blackwell, Oxford.

12. Bottomore, Tom, (1983): *A Dictionary of Marxist Thought*, London.
13. Bottomore, Tom, (1965): *Classes in Modern Society*, Allen & Unwin, London.
14. Bottomore, Tom (ed.)., (1981): *Modern Interpretations of Marx*, Oxford: Basil Blackwell.
15. Bottomore, Tom, and Goode, J.Patrick (eds)., (1983): *Readings in Marxist Sociology.*, Clarendon Press, Oxford.
16. Braverman, Harry, (1974), *Labour and Monopoly Capitalism The Degradation of Work in the Twentieth Century*, Monthly Review Press. New York.
17. Bukharin, Nicholai, (1929): *Imperialism and World Economy*, New York.
18. Carver, Terrell, (1975): *Karl Marx Texts on Method*, Basil Blackwell, Oxford.
19. Dickson, D., (1974): *Alternative Technology and the Politics of Technical Change*, Fontana.
20. Dobb, Maurice, (1986): *On Economic Theory and Socialism. Collected Papers*, Universal Book Stall, New Delhi.
21. Dreze, Jean and Sen Amartya, (1993): *Hunger and Public Action*, *Oxford* University Press, Delhi.
22. Emmanuel, A., (1972): *Unequal Exchange: A Study of Imperialism of Trade*, London.
23. Frank, Andre Gunder, (1967): *Capitalism and Underdevelopment in* Latin American: Historical Studies of *Chile and Brazil*, New York.
24. Furtado, C., (1964): *Development and Underdevelopment*, University of California Press
25. Furtado, Celso, (1963): *The Economic Growth of Brazil: A Survey from Colonial to Modern Times*, Berkeley.
26. Guha.S., Ashok, (1981): *An Evolutionary View of Economic Growth*, Clarendon Press, Oxford.
27. Hayek, F.A. (ed.)., (1954): *Capitalism and the Historians*, Routledge & Kegan Paul, London.
28. Hirschman, A., (1958), *The Strategy of Economic Development*, Yale University Press.
29. Johnson, H.G., (1958): *International Trade and Economic Growth*, Allen & Unwin.

30. Kaldor, N., (1957): "A Model of Economic Growth", *Economic Journal*, Vol. LXVH,'
31. Knights, David and Willmott, Hugh, (Ed.), (1988): *New Technology and the Labour Process,* The Macmillan Press.
32. Kuznet, Simon, (1971): *Economic Growth of Nations*, Cambridge.
33. Lal, Sanjay, (1975): Dependency a Useful Concept in Analysing Underdevelopment?", *World Development*, Vol. III, Nos. 11-12.
34. Lawson, Rowena, (1984): *Economics of Fisheries Development*, Frances Printer, (Publishers), London.
35. Lefeber, Louis, (1974): "On the Paradigm of Economic Development", *World Development*, Vol. 2, No. 1.
36. Lenin, V.I., (1964): "Imperialism, the Highest Stage of Capitalism", in *Collected Works*, Vol. 22, Progress Publishers, Moscow
37. Lewis, W.A., (1955): The Theory of Economic Growth, Alien & Unwin.
38. Luxemburg, Rosa, (1964): *The Accumulation of Capital,* New York.
39. Mandel, Ernest, (1974): *Late Capitalism*, New Left Books, London.
40. Mansfield, E., (1969): *The Economics of Technological Change*, Longman
41. Marx, Karl, (1978): Capital, 3 Volumes, Progress Publishers Moscow.
42. Meade, J.E., (1961): *A Neo-Classical Theory of Economic Growth*, Unwin University Books.
43. Meier, G., (1968): *The International Economics of Development*, Theory and Policy, Harper &Row.
44. Myrdal, G., (1957): *Economic Theory and Underdeveloped Regions*, Duckworth.
45. Myrdal, Gunnar, (1968): *Asian Drama An Inquiry into the Poverty of Nations,* Vol. I, Middlesex.
46. Nurske, R., (1953): *Problems of Capital Formation of Underdeveloped Countries,* Blackwell.
47. Poulantzas, Nicos, (1973): *Political Power and Social Classes*, New Left Books, London.
48. Ranabir, Samadar, (1994): *Workers and Automation, The Impact of New Technology in the News Paper* Industry, Sage Publications, New Delhi.
49. Robinson, Joan, (1974): *An Essay on Marxian Economics*, Macmillan.

50. Robinson, Joan, (1979): *Aspects of Development and Underdevelopment,* Cambridge University Press, London

51. Robinson, Joan, (1982): *An Essay on Marxian Economics* The Macmillan Press Ltd.

52. Rosenberg, N., (1969): "The Direction of Technological Change: Inducement Mechanisms and Focussing Devices", Economic Development and *Cultural Change* Vol. 18.

53. Rosenberg, N., (1963): "Technological Change in the Machine Tool Industry, 1840-1910", Journal of *Economic History,* Vol. XXIII.

54. Rostow, W.W., (1960): *The Stages of Economic Growth*, Cambridge.

55. Roy, M.N., (1922): *India in Transition,* Geneva.

56. Rudra, Ashok, (1988): *Some Problems of Marx's Theory of History*, Calcutta.

57. Sau, Ranjit, (1981): *India's Economic Development Aspects of Class Relations*, Orient Longman.

58. Sau, Ranjit, (1978): *Unequal Exchange, Imperialism and Underdevelopment*, Delhi.

59. Schumpeter, J.A., (1934): *The Theory of Economic Development,* Harvard University Press. Cambridge.

60. Schumpeter, J.A. (1976): *Capitalism, Socialism and Democracy* Allen & Unwin, London.

61. Sen, A.K., (1966): "Peasants and Dualism with or without Surplus Labour", Journal of *Political Economy*, Vol.

62. Sen, A.K., (1968): *Choice of Techniques*, Blackwell.

63. Sen, Anupam, (1982): *The State, Industrialization and Class Formations in India, A Neo-Marxist Perspective on Colonialism, Underdevelopment and Development*, *Routledge* & Kegal Paul, London.

64. Sen, Ashok, (1984): "The Transition from Feudalism to Capitalism: Review of Political Economy", *Ecnomic and Political* Weekly, July.

65. Sen, A.K., (1975): *Employment, Technology and Development*, Clarendon Press.

66. Sen, Amartya, (1981): *Poverty and Famines*, London.

67. Singer, H., (1969): 'The Mechanisms of Economic Development', *Indian Economic Review*, Vol. I.

68. Smith, Adam, (1776): *The Wealth of Nations*, London.

69. Stewart, Frances, (1977): *Technology and Underdevelopment*, London.

70. Streeten, P.P., (1972): "Technology Gaps between Rich and Poor Countries", *The Scottish Journal of Political Economy*, Vol. XDC

71. Streeten, Paul, (1972): *The Frontiers of Development Studies*, London.

72. Sweezy, Paul, (1946): *The Theory of Capitalist Development*, Oxford University Press, New York.

73. Sweezy, Paul, (1972) : *Modern capitalism and Other Essays*, Monthly Review Press, New York and London.

74. Tawney, R.H. (1952): *Equality*, Allen & Unwin, London.

75. Zimbalist, A., (ed.), (1979): *Case Studies in the Labour Process* Monthly Review Press, London.

Index